GPW WILLY'S
1/4 TON MILITARY TRUCK MANUAL

TM 10-1513

Maintenance Manual May 1942 Change 1

edited by
Brian Greul

The GPW Model MB, commonly referred to as the Jeep is probably the most ubiquitous American military vehicle ever produced. Simple, rugged, and capable of hard work it began as a World War II vehicle and descendant vehicles are still produced today as passenger vehicles.

This book is intended to support enthusiasts and their restoration efforts by providing a professionally printed, 8.5x11 compilation of the key manuals for this vehicle.

Every effort has been made to faithfully reproduce the document while cleaning up the pages to make them usable to you the reader. However, we are dealing with original works that have been electronically preserved from nearly 80 years ago. There are a number of artifacts in the source documents. Your understanding is appreciated.

An 8.5x11 3 hole punched loose leaf copy may be purchased for your 3 ring binder. Email books@ocotillopress.com for current information.

Should you have suggestions or feedback on ways to improve this book please send email to Books@OcotilloPress.com

---

Edited 2021 Ocotillo Press
ISBN 978-1-954285-14-9

Printed in the United States of America

Ocotillo Press
Houston, TX 77017
Books@OcotilloPress.com

Disclaimer: The user of this book is responsible for following safe and lawful practices at all times. The publisher assumes no responsibility for the use of the content of this book. The publisher has made an effort to ensure that the text is complete and properly typeset, however omissions, errors, and other issues may exist that the publisher is unaware of.

MAINTENANCE MANUAL

FOR

WILLYS TRUCK

¼ TON 4 x 4

BUILT FOR
U. S. GOVERNMENT

MODE-L MB

Contract Number

W-398-qm-11423

U. S. A .. Reg. Numbers
2073506 to 2078606

✱ ✱ ✱

Parts are designated in this book under both
Ford and Willys part numbers since all parts are interchangeable for
vehicles produced by Ford Motor Company.

Contract **W-398-qm-11424** Model GPW

U. S. A. Registration Numbers
20l00000Sto20163145S

TM-10-1349

WILLYS-OVERLAND
MOTORS, INC.
TOLEDO,OHIO,U. S.A.

TM-10-1513

. CHANGE NO. I MAY 15,
1942

TM-10-1513
CHANGE NO. I MAY 15,
1942

INDEX

•section Page

	•section	Page
Unloading	0-1	4
Drivers Inatruction•	0-2	5
Lubrlection and Inspection	0-3	9
Englne	0100	17
Clutch	0200	37
Fuel System	0300	42
Exhaust Sytem	0400	52
Coolinb	0500	53
Electrical	0600	57
		73
Transfer Case	0800	79
Propeller Sh.rt and Universal Joints	0900	83
Front Axle	1000	86
Rear Axle	1100	93
Brakes	1200	101
Wheels, hubs and Drums	1300	108
Steering	1400	Ill
Frame	1500	119
Springs and Shock Absorbers	1600	121
Body	1800	125
Tools	2300	126

• These numbers refer-to Parts Group
Classification In Parts List.

FOREWORD

This Motor Vehicle has been thoroughly tested and inspected. Like any other piece of machinery, to maintain it in proper operating condition, it should be lubricated at the time specified using the proper grades of oil and grease. All working parts as well as oil holes should be kept clean and free from dirt and grit. This vehicle should periodically have a systematic inspection.

All parts in this vehicle are completely interchangeable with those manufactured by Ford Motor Company under the contract listed on the preceding page. Both Ford and Willys part numbers are therefore listed under the illustrations showing views of the various assemblies. These part numbers should be used only for the purpose of identifying parts as they are mentioned in the text, and, the accuracy of the part number should be verified by referring to the parts book when placing orders for parts.

In the following pages we have described how to take care of this unit and handle it in such a way that it will give maximum service and dependable performance.

In the forepart of this Manual will be found complete instructions relative to conditioning the unit for Service, Driver's Instructions, Lubrication and Inspection.

In the Maintenance and Repair Section will be found instructions which will enable one to make proper adjustments and repairs.

See Index on preceding page; bend back edge of pages to find Section desired.

Read and follow these instructions carefully.

WILLYS-OVERLAND MOTORS, INC.

WILLYS MODEL "MB" 1/4-TON 4 x 4 GOVERNMENT TRUCK

GENERAL DATA

ENGINE

Type..Gasoline
Number of Cylinders...............................4
Bore..3⅛"
Stroke..4⅜"
Piston Displacement...............................134.2 cu. in.
Compression Ratio.................................6.48-1
Horsepower—S.A.E..................................15.6
Horsepower ⌠ Actual...............................60
        ⌡ Revolutions per minute.................4000
Torque ⌠ Maximum Lbs.-Ft..........................105
     ⌡ Revolutions per Minute.................2000
Wheelbase...80"
Tread.................................48¼"-with combat wheels 49"
Overall Width.....................................62"
Overall Length....................................132¾"
Overall Height—Normal Load
    To top of cowl................................40"
    To top of steering wheel......................51¼"
    Top up..69¾"
Weight—Maximum Pay Load...........................800 lbs.
    Maximum Trailed Load..........................1000 lbs.
    Shipping (Less water, fuel and chains)........2125 lbs.
    Road..2315 lbs.
    Gross...3125 lbs.

CAPACITIES

	U.S.	Imperial	Metric
Fuel Tank (Gals.)	15	12½	56.78 liters
Engine Crankcase-Refill (Qts.)	4	3½	3.78 "
Cooling System (Qts.)	11	9¼	10.41 "
Transmission (Pts.)	2	1¾	.95 "
Transfer Case (Pts.)	3	2½	1.42 "
Front Axle Differential (Pts.)	2½	2	1.18 "
Rear Axle Differential (Pts.)	2½	2	1.18 "
Oil Bath Air Cleaner (Pts.)	1¼	1	.71 "
Brake System Brake Fluid (Pts.)	¾	¾	.36 "

See Lubrication Chart, Page 12

LAMP BULBS Mazda

Head Lamp (Sealed Beam type)....................2400
    Upper Beam..................................45 Watts
    Lower Beam..................................45 Watts
Blackout Lamp Bulb (1)..........................3 Cp. SC 63
Left Tail Lamp Bulb (1).........................21-3 Cp. DC 1154
Left Tail Lamp Bulb (1).........................3 Cp. SC 63
Right Tail Lamp Bulbs (2).......................3 Cp. SC 63
Instrument Lamp Bulb (2)........................1.5 Cp. SC 51
Fuse (Thermal Type)—On Light Switch-30 Amperes

IDENTIFICATION

Chassis Serial Number located on inside of frame at

Engine Number located on right side of cylinder block

UNLOADING INSTRUCTIONS

Spot freight car along side of the unloading platform. Open freight car door and make visual inspection of vehicles for damage, loose blocking and shortages, due to rough handling or pilferage while vehicles were in transit. If any evidence of carrier's responsibility, the railroad representative should inspect shipment and note it on Bill of Lading.

Vehicles are shipped from one to six in a freight car, therefore, the manner varies in which the vehicles are anchored in the car. Where shipment does not exceed two vehicles per freight car, the regular 36 foot box car is used. Where three or more vehicles are shipped an "Evans" or "Channel" automobile freight car is used. These freight cars are equipped with upper deck platforms operated by chain falls and have anchor chains in flooring; to operate follow printed instructions on inside wall at controls.

One or Two Vehicles per Car

The vehicles are anchored to floor with grooved blocks spiked to the floor at front and rear of each wheel. Spring rebound straps are anchored to front end of front springs and rear end of rear springs and spiked to the floor.

To remove vehicles from car, use a crow bar to pry loose wheel blocks and straps from floor. Remove bolt in spring rebound strap at springs and remove straps.

Roll one vehicle to end of car, then jack or lift the other vehicle so it can be removed through door to platform, then remove second vehicle, and check all items listed in Tool and Accessory list.

Three Vehicles per Car

Where three vehicles are shipped, the two end vehicles are fastened at front end with car equipment anchor chains. The rear wheels have grooved blocks spiked to the floor. Spring rebound straps at end of rear springs are also spiked to the floor.

The center vehicle is anchored at the ends of front and rear axles with car equipment chains. Spring rebound straps at end of front and rear springs are spiked to the floor.

To remove vehicles first remove all wooden blocks, spring rebound straps and anchor chains from the three vehicles. Run end vehicles to extreme ends of freight car; jack or lift center vehicle so it can be rolled through door to platform. Repeat this operation to remove other two vehicles.

Four Vehicles per Car

Where four vehicles are shipped, one is decked and three anchored to the floor the same as in three vehicle shipment.

To remove vehicles, first remove anchor chains and wooden blocks from the three vehicles on floor and remove vehicles to platform. Follow instructions printed on inside of freight car at controls in ends of car for lowering Deck platform. Lower platform and remove anchor chains, then remove vehicle.

Five Vehicles per Car

Two vehicles are decked and three anchored to flooring in same manner as four to a car.

The removal of vehicles should be in the same sequence as outlined under three and four car shipment.

Six Vehicles per Car

Where six vehicles are shipped, two are decked and four are anchored to the floor.

The two end vehicles are fastened at front ends with anchor chains, the rear end of vehicles are anchored with grooved blocks and spring rebound straps spiked to the floor.

The two center vehicles are fastened in the opposite manner, rear ends with anchor chains and front ends with wheel grooved blocks and spring rebound straps spiked to the floor.

To remove vehicles remove wheel blocks, spring rebound straps and anchor chains. Roll end cars and one center car to end of freight car, jack or lift other center vehicle so it can be removed to platform, then remove other three.

Lower one decked vehicle by chain falls, following instructions printed on wall. Then remove second decked vehicle in same manner.

PRE-OPERATION INSTRUCTIONS

All vehicles are carefully tested and inspected before leaving the factory, however, while in transit and unloading some things may happen which will require attention before putting vehicle into Service. We therefore suggest checking the following items before operating vehicle.

1. Fill radiator and check all connections for water leaks.
2. Check oil in engine, transmission, transfer case, front and rear differential housings.
3. Fill gasoline tank and check full system for leaks.
4. Check battery fluid level.
5. Check terminal connections at battery, generator, voltage control, starter, distributor and spark plugs.
6. Check operation of lights and horn.
7. Check brake fluid level in master cylinder and check connections for leaks or damage.
8. Check steering connections and front wheel alignment.
9. Check tire pressure, inflate to 30 lbs.
10. Check hand brake operation.
11. Check cylinder head screws and nuts.

DRIVER'S INSTRUCTIONS

This vehicle should not be driven faster than 40 miles an hour for the first 100 miles nor more than 50 miles an hour from 100 to 500 miles. If the vehicle is operated at excessive speeds while new, the closely fitted parts may possibly become overheated, resulting in serious damage to mechanical units. Never race the Engine while making adjustments or when vehicle is standing idle.

FIG. 1—CONTROLS

It is very important that the driver of this vehicle be thoroughly familiar with the various Controls and their proper use. The most experienced driver should study the Controls because there are a number which are not ordinarily found on standard vehicles.

Illustrations show the controls, instruments and instruction plates; in the following paragraphs we refer to these illustrations by the key numbers so the reader may easily follow the instructions.

Ignition Switch—No. 9, Fig. 1

Is operated by a key, turning key to right (clockwise) closes the ignition circuit. Turning key to left (counter clockwise) opens the ignition circuit and shuts off the engine.

Light Switch—No. 6, Fig. 1

The light switch is the push-pull type with safety lock.

This switch controls the entire lighting system including the instrument panel lights and stop lights.

When the control knob is pulled out to the first position, the blackout lamp circuit is closed—which consists of two blackout lamps, stop and tail lamps.

To obtain bright lights, push in lockout control button on left of switch and pull out control knob to second position. This closes entire bright light circuit, which consists of two head lamps—instrument panel lamps, stop and tail lamps.

CAUTION: When driving during the day press in lockout control button and pull Control Knob out to the last or Stop Light position to cause regular Stop Light to operate.

Panel Light Switch—No. 12, Fig. 1.

The Panel Light switch controls the Panel Lights when the main Light Switch is in Service (bright light) position, otherwise the Panel Lights do not operate.

Head Lamp Beam Control Switch—No. 8, Fig. 1

Pressing and releasing the button of the selector foot switch with the left foot alternately changes the headlight beam from high to low.

Starter Switch—No. 23, Fig. 1

Toe board mounted to the right of the accelerator; pushing button down closes starter circuit and causes starter to crank engine—release the button as soon as the engine starts.

Hand Throttle—No. 10, Fig. 1

Pulling control button out opens carburetor throttle valve and increases engine speed.

Carburetor Choke Control—No. 7, Fig. 1
Pulling control button out closes choke valve in carburetor to enrich gas mixture for starting the engine when cold, and opens throttle valve slightly for faster idle speed.

Oil Gauge—No. 15, Fig. 1
The instrument panel oil gauge indicates oil pressure delivered to camshaft, crankshaft, timing chain and connecting rod bearings when engine is running.

Proper registration should be not below 10 on idle nor more than 80 at speeds above 10 miles per hour.

This gauge does not indicate the amount of oil in crankcase.

Ammeter—No. 20, Fig. 1
The ammeter is used to indicate when the generator is charging the battery. It also indicates the amount of current being consumed.

If the ammeter shows discharge at all times, the cause should be immediately investigated and corrected, otherwise the wiring may be damaged and battery discharged.

Fuel Gauge—No. 13, Fig. 1
The fuel gauge registers the amount of fuel in the fuel tank when ignition switch is turned on. The dial graduations are for—empty, ¼, ½, ¾, and full.

Temperature Indicator—No. 19, Fig. 1
This is a thermal type gauge and registers the temperature of the liquid in the cooling system. The operator should watch this instrument closely.

The normal operating temperature is indicated when hand stands between 160 and 185. The driver should immediately investigate the cause if temperature becomes excessive. Continuous operation of an overheated engine will cause serious damage.

Never fill cooling system with cold water when engine is overheated.

Speedometer—No. 17, Fig. 1
The Speedometer indicates the speed at which vehicle is being driven. The Odometer (in upper part of speedometer face) registers the total number of miles the vehicle has been driven.

A trip mileage indicator (in lower part of speedometer face) gives miles covered on any trip. It

can be reset by turning a knurled control shaft extending through the rear of the speedometer.

Nomenclature Plate (Name Plate) —Fig. 2
The nomenclature plate identifies the vehicle and gives the manufacturer's model and serial number, date of delivery, recommended fuel and lubricating oil. Service publication numbers are also given for reference. (When ordering parts be sure to give serial number). See No. 26, Fig. 1.

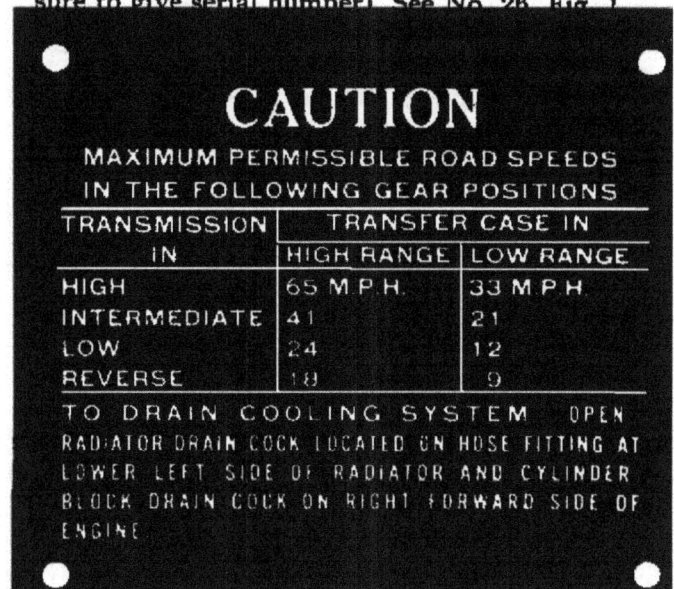

FIG. 3—CAUTION PLATE

Caution Plate—Fig. 3 & No. 29, Fig. 1
Covers maximum permissible road speeds in different gear positions and gives instructions relative to complete draining of the cooling system.

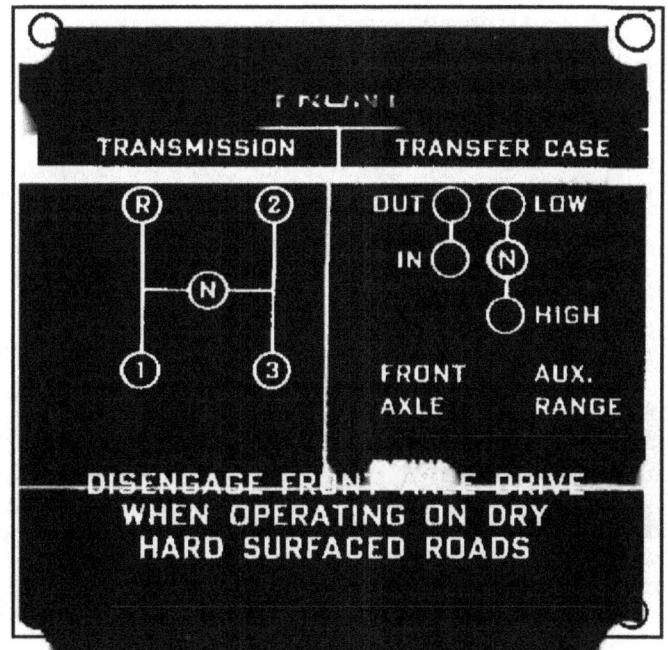

FIG. 4—SHIFT PLATE

Transfer Case Shifting Instructions—Fig. 4
This diagram gives relative position of shifting levers for front axle drive, low and high gear ratios.

On hard surface and flat roads disengage front axle drive by placing center shift lever, (front axle drive) in forward position. No. 21 & No. 25, Fig. 1.

The right hand lever (third from driver) controls transfer case gear ratio—low or high. No. 22, Fig. 1. The low gear ratio can only be used when front axle drive lever is in the rear position to engage front wheel drive.

Proper position for disengaging axles to use power take-off unit is shown as "N" in Fig. 4.

Clutch Pedal—No. 11, Fig. 1

The clutch pedal is used to disengage the engine power from transmission when shifting gears. Driving with the foot on pedal will cause excessive wear of clutch facings and release bearing. There should be ¾" free pedal travel before clutch starts to disengage.

Brake Pedal—No. 14, Fig. 1

Depressing the pedal applies the hydraulic brakes at all four wheels. Avoid driving with foot on brake pedal, as brakes will be partially applied and cause unnecessary wear of brake linings requiring early adjustment.

Hand Brake Lever—No. 24, Fig. 1

By pulling out on brake handle the external contracting brake at the transmission on rear propeller shaft is applied mechanically. Whenever vehicle is parked, the lever should be pulled out as far as possible. Before moving vehicle be sure lever is released.

Accelerator—No. 16, Fig. 1

The accelerator is foot operated and is used to govern the engine speed under ordinary driving conditions.

Transmission Gearshift Lever—No. 18, Fig. 1

This lever is used to shift the gears in the transmission. There are four positions in the movement of the lever in changing the gears in transmission. See diagram for lever location in different gears. Fig. 4.

Horn Button—No. 2, Fig. 1

Pressing on button closes circuit in horn wiring and causes horn to sound.

Steering Knuckle Oil Seal—Fig. 11, Page 90

When parking during cold, wet weather swing the front wheels from right to left to wipe away moisture adhering to the front axle universal joint housings and oil seals. This will prevent freezing with resulting damage to the oil seal felts. When the car is stored for any period the front axle universal joint housings should be coated with light grease to prevent rusting.

Fuel Tank

The fuel tank is located under Drivers seat. To fill tank raise seat cushion and remove filler cap. See Capacity Chart, Page 3.

Draining Radiator

When draining radiator be sure to open the drain cock on the right forward side of cylinder block as well as at the bottom of the radiator outlet. Remove the radiator cap to break any vacuum and thoroughly drain system. See Caution Plate, Fig. 3.

OPERATING INSTRUCTIONS

Before any trip the following inspections should be made before starting engine.

1. Check the oil level in crankcase, See "Lubrication" Section. Remove oil level indicator located in oil filler pipe on right side of engine, wipe off clean. Insert indicator in filler pipe to full depth. Remove indicator and note position of oil film, if below the full level add sufficient oil to bring to full mark.

2. Remove radiator filler cap and note water level. Should be up within ½" below filler neck. Check all hose connections for leaks also fan belt tension.

3. Check all lights and signal devices. Note condition of tires and see that they are properly inflated to 30 lbs.

4. DRIVING THROUGH WATER. See that Cap is on front drain hole under fuel tank so as to keep out stones and dirt. An extra cap is provided for the rear drain hole and this cap should be kept in the glove compartment so it will be readily available for use when fording small streams. Before driving through streams or deep puddles of water, INSTALL THE CAP ON THE REAR DRAIN HOLE. Remove this cap and return it to the glove compartment after passing through the water.

5. When there is a possibility of water being thrown over the engine by fan action in crossing streams, pull up on handle of the generator brace, then remove the fan belt. This will stop the fan. As soon as possible the belt should be replaced, then pull out on the generator. The generator will lock in place by spring action of the brace.

STARTING THE ENGINE

1. Transmission gearshift lever must be in neutral position. See Fig. 4.

2. Pull out hand throttle about ¾" to 1".

3. Pull out choke button all the way to obtain proper fuel and air mixture for starting, No. 7, Fig. 1. Choking is not necessary when engine is warm.

4. Insert key in ignition switch and turn to right.

5. Disengage clutch by depressing pedal, holding down till engine starts, No. 11, Fig. 1.

6. Step on starter button No. 23, Fig. 1, to crank engine. Release button when engine starts.

7. Push in on choke button and adjust hand throttle to obtain proper idling speed. When engine is cold, it is advisable to leave choke button pulled out about 1". As engine warms up, push choke button all the way in.

STARTING VEHICLE

(For day time driving, turn on Stop Light; See Light Switch, Page 02-5.)

1. Push clutch pedal down to disengage clutch No. 11, Fig. 1.

2. Shift transfer case (center hand lever) in forward position (front axle disengaged) No. 21, Fig. 1, right hand lever No. 22, Fig. 1 in rear position (high gear ratio).

3. Move transmission shift lever toward driver and back for first gear. Fig. 4.

4. Release hand brake, No. 24, Fig. 1, increase engine speed with accelerator by gradually pressing down on accelerator treadle No. 16, and slowly release clutch pedal, No. 11, increasing engine speed as load is picked up and vehicle starts to move.

5. As vehicle speed increases, release accelerator pedal and depress clutch pedal, move gearshift lever to neutral then in to second gear.

Press on accelerator and release clutch pedal slowly. Repeat these operations until transmission is in high gear.

SHIFTING GEARS IN TRANSFER CASE

Instructions for shifting gears in transfer case and engagement of the front axle drive are as follows: No. 21, Fig. 1 and Fig. 4.

1. Transfer case may be operated in either high or low speed range when front axle drive is ENGAGED.
2. The transfer case can be operated only in "High" (direct drive) when front axle drive is DISENGAGED.
3. To engage front axle drive, depress clutch pedal, release accelerator and move center shift lever to rear position, No. 21, Fig. 1.
4. To disengage front axle drive, release accelerator and shift lever to forward position.
5. Shifting from high to low gear should not be attempted except when the vehicle is being

FIG. 5—CHAIN TOW

operated at low speeds or at a standstill. The front axle drive must be engaged for this shift. Release accelerator and depress clutch pedal—move center shift lever to rear position, engaging front wheel drive, No. 21, Fig. 1, then move right hand shift lever, No. 22, to forward position.
6. Shifting from low to high gear may be accomplished at any time, regardless of vehicle speed. Release accelerator and depress clutch pedal—shift right hand lever into rear position.

SHIFTING TO LOWER SPEED IN TRANSMISSION

The transmission gears should always be shifted to the next lower speed before engine begins to labor or before vehicle speed is reduced appreciably. Shifting to lower speed is accomplished as follows:

1. Depress clutch pedal quickly, increase engine speed and shift to next lower gear, release clutch slowly and accelerate.

It is advisable to use the same transmission gear going down a long hill as would be required to climb the same hill.

STOPPING THE VEHICLE

1. Remove foot from accelerator pedal and apply brakes by pressing down on brake pedal. No. 14, Fig. 1.
2. When vehicle speed has been reduced to idling engine speed, disengage clutch and move transmission shift lever, No. 18, Fig. 1 to neutral position. See Fig. 4.
3. When vehicle has come to a complete stop apply hand brake, No. 24, Fig. 1, and release clutch and brake pedals.

SHIFTING INTO REVERSE

Before attempting to shift into reverse the vehicle must be brought to a complete stop.

1. Push clutch pedal down to disengage clutch.
2. Shift transmission lever to the left and forward toward instrument board. Fig. 4.
3. Release clutch pedal slowly and accelerate as load is picked up.

FIG. 6—ROPE TOW

TOWING VEHICLE

When necessary to tow vehicle the tow chain, rope or cable, should be attached to the front bumper bar and frame side rail gusset, Fig. 5 and 6.

Loop chain or rope over top of bumper and frame gusset bringing it up across face of bumper and back on opposite side of frame, then hook or tie. Do not tow from the middle of the bumper.

FIRE EXTINGUISHER

The fire extinguisher is mounted on left side cowl panel Fig. 1. To remove pull outward on clamp release lever.

To operate extinguisher, hold body in one hand and with the other turn handle to left ¼ turn which releases plunger lock. Use pumping action to force fluid on fire.

Read instructions on fire extinguisher plate.

GENERAL LUBRICATION

Lubrication of any vehicle is important to prevent damage to moving parts. Because all moving parts are not subjected to the same operating conditions, the lubricants specified are those which most nearly meet the requirements of the parts involved. In some places excessive heat or cold is a problem to overcome, in others it is extreme pressure, water, sand or grit. The type of operating surfaces must also be taken into consideration as parts rotate or oscillate on various types of bearings. Each of the above conditions in construction make necessary the application of the specified lubricant.

Lubricants should be applied regularly to secure maximum useful service from the vehicle. It is of equal importance that not only the proper grade of lubricant be used but that it be applied in accordance with a definite schedule.

The chart in this section should be referred to for instructions on mileage of application, grade and quantity of lubricant required for all parts of the vehicle. A more detailed account of certain phases of lubrication is given in the following paragraphs.

ENGINE

Lubrication of the engine is accomplished by means of a force-feed continuous circulating system. This is effected by means of a planetary gear type pump located externally on the left side of the engine, and is driven by a spiral gear on the camshaft.

The oil is drawn into the circulating system through a floating oil intake. The floating intake does not permit water or dirt to circulate, which may have accumulated in the bottom of the oil pan, because the oil is drawn horizontally from the top surface. Oil pressure is maintained under all driving and climatic conditions.

Oil is forced to the crankshaft and camshaft bearings through drilled passages in the cylinder block and then to the connecting rod bearings through drilled passages in the crankshaft. A drilled passage in the crankshaft, from the front bearing to holes in the crankshaft sprocket provides positive lubrication for the timing chain. Direct spray from connecting rod bearings lubricates the cylinder walls, pistons, piston pins and the valve mechanism.

The pressure under which the oil is forced to the bearings is controlled by a pressure regulator or relief valve, located in the cover of the oil pump. The valve is set to relieve at an indicated pressure of 75 lbs. at a car speed of approximately 30 miles per hour, with warm oil, assuring ample lubrication at all speeds. An oil pressure gauge is mounted in the instrument panel, and indicates the pressure being supplied. Failure of the gauge to register may indicate absence of oil or leakage and the engine should be stopped immediately.

If there is plenty of oil in the reservoir, the oiling system should be carefully checked before starting the engine.

For capacity of the oiling system see Page 3. Care should be taken to replenish the supply when the oil level indicator, which is combined with the oil filler cap located in the oil filler pipe, shows the oil below the full mark, No. 6, Fig. 1. Fresh oil should be poured into the reservoir through the filler pipe sufficiently to bring level to full mark.

WHEN TO CHANGE CRANKCASE OIL

When the vehicle leaves the factory the crankcase is filled to the correct level with oil of the proper viscosity for the "break-in" period. (Decked vehicles in freight cars have engine oil drained and five quarts of oil in cans in freight car for each vehicle.)

At 500 miles and 1500 miles, then every 2500 miles thereafter completely drain the oil by removing the drain plug, in the lower left side of the oil pan and refill with 4 quarts of fresh lubricant, in accordance with specifications.

To insure continuation of best performance and long engine life, it is necessary to change the crankcase oil whenever it becomes diluted or contaminated with harmful foreign materials. Under the adverse driving conditions described in the following paragraphs, it may become necessary to drain the crankcase oil more frequently.

Vehicles operated in extremely dusty country, should have the oil drained both winter and summer, at 1,000 mile intervals or oftener, and extra precaution should also be taken to keep the carburetor airfilter clean and supplied with oil. The frequency of cleaning the Carburetor Oil Bath Air Cleaner depends upon severity of dust conditions and no definite refill periods can be recommended.

Thinning of the oil by unburned fuel leaking by the piston rings and mixing with the oil, is known as crankcase dilution.

Leakage of fuel into the oil pan mostly occurs during the "Warming-Up" period, when the fuel is not thoroughly vaporized and burned.

Short runs in cold weather do not permit thorough warming up of the engine and water may accumulate in the crankcase from condensation of moisture from piston blow-by.

Practically all present-day engine fuels contain a small amount of sulphur which, in the state in which it is found, is harmless; but this sulphur on burning, forms certain gases, a small portion of which is likely to leak past the pistons and rings and reacting with water when present in the crankcase forms sulphurous acid. As long as the gases and the internal walls of the crankcase are hot enough to keep water vapor from condensing no harm will result, but when an engine is run in low temperatures moisture will collect and unite with the gases formed by combustion, thus acid will be formed and is likely to cause serious etching or pitting. This manifests itself by broken valve springs and excessively rapid wear on piston pins, crankshaft bearings and other parts of the engine.

In view of these conditions it is necessary to drain the crankcase oil at regular intervals. It is always advisable to drain the oil when the engine is warm. The benefit of draining is, to a large extent, lost if the crankcase is drained when the engine is cold because some of the foreign material will remain in the bottom of the oil pan and will not drain out readily with the oil.

At least once a year, preferably in the Spring, the oil pan and floating oil intake should be removed from the engine and thoroughly washed with cleaning solution.

Lubrication Chart Index

See Page 13 for Details

1—Spring Shackle (4)
2 hydraulic fittings
Pressure gun
Chassis grease #1

2—Spring Bolt (5)
1 hydraulic fitting
Pressure gun
Chassis grease #1

3—Tie Rod (2)
2 hydraulic fittings
Pressure gun
Chassis grease #1

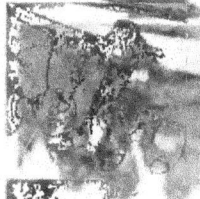

4—Drag Link
2 hydraulic fittings
Pressure gun
Chassis grease #1

6—Universal Joint
Needle Bearings (4)
1 hydraulic fitting in center
Pressure gun (hand) & adapter
Mineral oil gear lubricant

7—Slip Joint (2)
1 hydraulic fitting
Pressure gun
Chassis grease #1

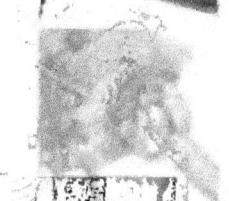

10—Lever Shaft
Transfer case shift
1 hydraulic fitting
Pressure gun
Chassis grease #1

10—Lever Shaft
Clutch and brake pedal
2 hydraulic fittings
Pressure gun
Chassis grease #1

10—Lever Shaft
Steering belcrank
1 hydraulic fitting
Pressure gun
Chassis grease #1

19—Wheel Bearings (4)
Remove and repack
Chassis grease #1

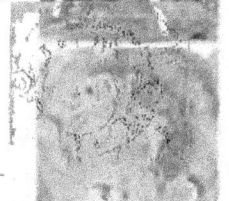

21—Linkage
All clevis pins
Oil can
Engine oil

22—Steering Gear Housing
1 plug
Pressure gun (hand)
Chassis grease #1

27—Front Axle Universal (2)
1 plug—fill to level
Pressure gun
Chassis grease #1

28—Transmission
1 fill and 1 drain plug
Pump
Mineral oil gear lubricant

29—Transfer Case
1 fill and 1 drain plug
Pump
Mineral oil gear lubricant

30—Axle Housing (2)
1 fill and 1 drain plug
Pump
Hypoid gear lubricant

34—Distributor
1 oiler—1 wick—1 post
Oil can
Engine oil
Grease cam

36—Starter
1 oil hole
Oil can
Engine oil

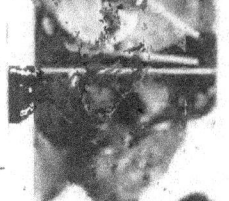

39—Crankcase
1 fill pipe and 1 drain plug
Oil can
Engine oil

40—Pintle Hook
2 points
Oil can
Engine oil

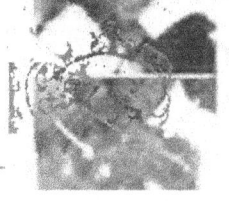

LUBRICATION CHART
¼ Ton 4 x 4 Chassis
Hydraulic Brakes

Make Willys
Model MB

1—Spring Shackle
2—Spring Bolt
3—Tie Rod
4—Drag Link
6—Universal Joint Needle Bearings
7—Slip Joint
10—Lever Shaft—Transfer Case Shift
10—Lever Shaft—Clutch Shaft and Brake Pedal
10—Lever Shaft—Steering Bel-crank
19—Wheel Bearings
21—Linkage
22—Steering Gear Housing
27—Front Axle Universal
28—Transmission
29—Transfer Case
30—Axle Housing
34—Distributor
36—Starter
39—Crankcase
40—Pintle Hook

TOOLS

Cleaning Rag
Adjustable Wrench
Square Shank Wrench, ⅜"
Wheel Bearing Nut Wrench
Screw Driver

INSTRUCTIONS

Clean and lubricate all points in the order indicated, except those which require disassembly. Clean all vents. Check and adjust level in housings. Disassemble as separately instructed. Drain as separately instructed. See page 13.

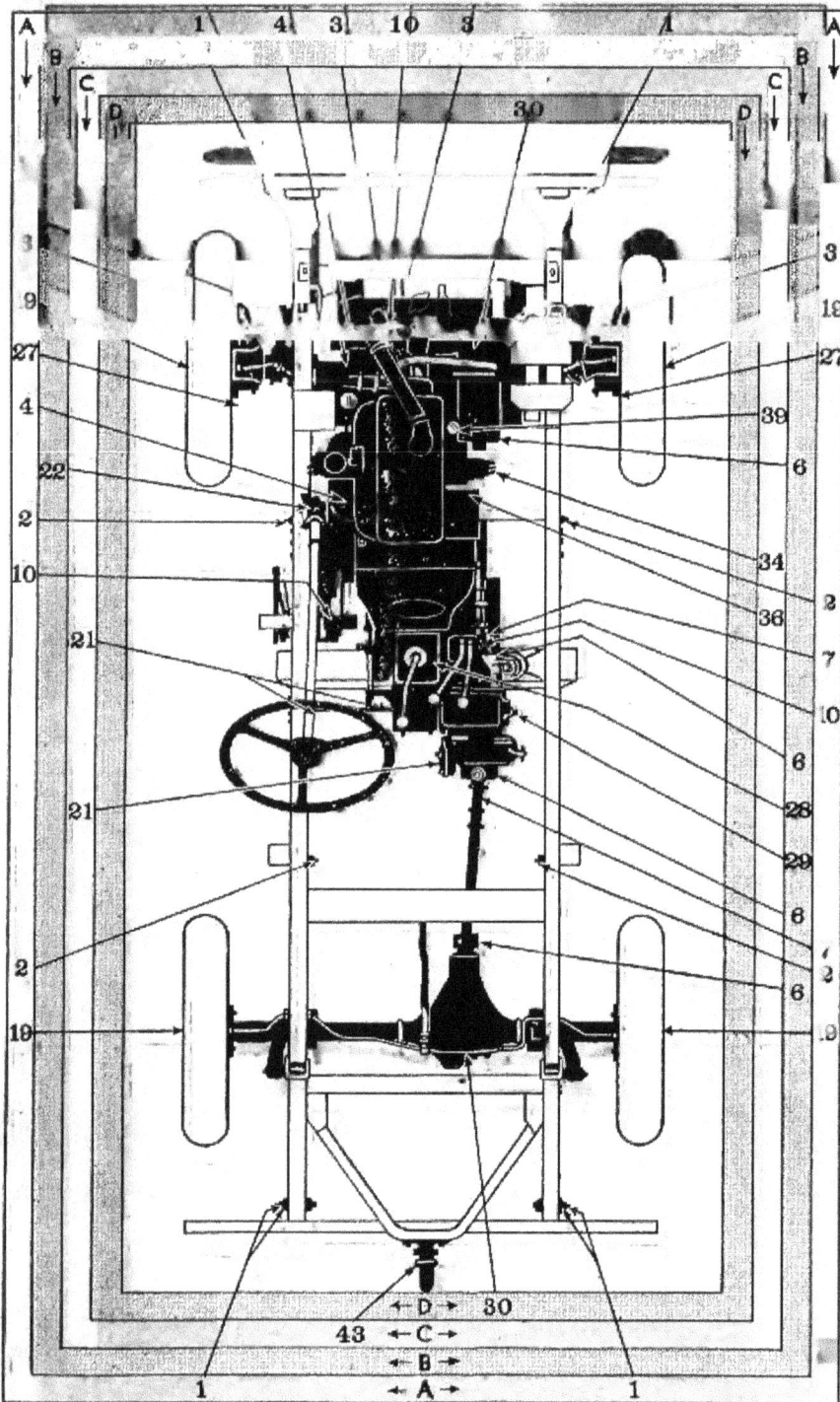

FIG. 1—CHASSIS

Frame A—Chassis Lubricant
Frame B—(Mineral Oil) Gear Lubricant.
Frame C—Engine Oil
Frame D—(Hypoid) Gear Lubricant.

Predominating Temperature

	Above 32° F.	Between 32° F. and 0° F.	Below 0° F.
Chassis Grease	#1	#1	#1
Gear Lubricant	90	90	80
Engine Oil	30	10W	10W plus 10% Kerosene

LUBRICATION CHART
U. S. GOVERNMENT

For Canadian and British Lubrication Specifications see reverse side of Index Page in front of Manual.

ITEM TO BE LUBRICATED See Page 10 & 11	HOW APPLIED	Capacity	Winter SAE	Winter Navy	Winter Army	Summer SAE	Summer Navy	Summer Army	MILES
Engine Crankcase (39)	Filler pipe R. side check level daily	Refill 4 qts	10W	1042		30	1065		2500
*Transmission Case (28)	Filler plug R. side—add Oil to level of plug	2 pts	90	1100		90	1100		6000
Transfer Case (29)	Filler plug—add oil to level of plug	3 pts	90	1100		90	1100		6000
Differential F. & R. (30)	Filler plug in cover—add Hypoid oil to level of plug	2½ pts	90EP		Fed. Spec. VVL 761 Class 2	90EP		Fed. Spec. VVL 761 Class 2	6000
Propeller Shaft Universal Joints F & R (6)	Fitting		140	1120		140	1120		1000
Air Cleaner	Remove Cover	1¼ pt	10W	1042		30	1065		2000
Front Axle Shaft Universal Joint & Steering Knuckle Bearings (27)	Filler plug outer casing	½ lb	NLGI No. 1		NLGI No. 1	NLGI No. 1		NLGI No. 1	1000
F & R Wheel Bearings (19)	Remove and Repack		NLGI No. 1		NLGI No. 1	NLGI No. 1		NLGI No. 1	6000
Steering Gear Housing (22)	Remove Plug	6½ oz	NLGI No. 1		NLGI No. 1	NLGI No. 1		NLGI No. 1	1000
Steering Bell Crank (10)	Fitting				Lubricate with NLGI No. 1 Summer & Winter				1000
Steering Tie Rods (3)	Fitting each end								
Steering Connecting Rod (4)	Fitting each end								
Spring Shackles F & R (1)	Fittings 8								
Spring Pivot Bolts F & R (2)	Fittings 2								
Clutch & Brake Pedal Shaft (10)	Fittings 5								
Propeller Shaft Slip Joints (7)	Fitting 1 each								
Starter Front (36)	Oil Hole	5 Drops	10W	1042		30	1065		1000
Distributor (34)	Oil Cup on side	5 Drops	10W	1042		30	1065		1000
Distributor Shaft Wick (34)	Oil Can	1 Drop	10W	1042		30	1065		2500
Distributor Arm Pivot (34)	Oil Can	1 Drop	10W	1042		30	1065		2500
Distributor Cam (34)	Wipe with grease		NLGI No. 1	NLGI No. 1		NLGI No. 1		NLGI No. 1	2500
Clevis Pins, Yokes & Cables (21)	Oil Can	5 Drops	10W	1042		30	1065		1000
Pintle Hook (40)	Oil Can	5 Drops	10W	1042		30	1065		1000
Hydraulic Brake System	Oil Can	¾ Pts.	Lockheed No. 21 Brake Fluid						

See Page 3 for Imperial and Metric qualities

*Remove Skid Plate to drain Transmission
Shock Absorbers Non-Refillable.

The following table shows at a glance the specification of oil to use in the engine according to temperature conditions:

	GRADE
Above 90° Fahr...	30
Not lower than 32° above zero Fahr....................	20 or 20W
As low as 10° above zero Fahr...........................	20W
As low as 10° below zero Fahr...........................	10W
Lower than 10° below zero Fahr.........................	10W plus 10% Kerosene

Always select an oil with a temperature range which agrees as closely as possible with the outdoor temperature range likely to be encountered. When the crankcase is drained and refilled, the oil should be chosen, not on the basis of the existing temperature at the time of change, but on the minimum temperatures that might reasonably be expected until time to change the oil again.

In warm weather, light oil tends to be used up a little faster than heavier oil; accordingly heavier oil is recommended for Summer use. In cold weather, however, it is important to use a light oil so that the engine can be started easily and to assure an adequate, early flow of oil to every part of the engine when first started and cold.

CHASSIS LUBRICATION

All hydraulic lubrication fittings indicated by No. 1-2-3-4 & 10 in Fig. 1 should be wiped clean and gone over with a compressor every 1,000 miles.

Make certain that each bearing surface is properly lubricated. All clevis pins, yokes and upper end of hand brake conduit should be oiled.

STEERING GEAR

Check level of lubricant in steering gear housing No. 22, Fig. 1, every 1,000 miles, keeping it filled at all times with NLGI No. 1 lubricant. Avoid the use of cup grease, graphite, white lead or heavy solidified oil.

Remove plug in steering housing and with a hand gun fill the housing slowly. When housing is full, replace the filler plug.

FAN AND WATER PUMP

The fan and water pump bearings are prelubricated and the lubricant lasts for the life of the bearings.

IGNITION DISTRIBUTOR

The oiler, on the distributor indicated by No. 34 in Fig. 1, should be lubricated every 1,000 miles with several drops of engine oil.

Every 2,500 miles when engine oil is changed apply a drop of light engine oil on the wick located in the top of the shaft which is accessible by removing the rotor arm. Also put a wipe of soft grease on the breaker arm cam, and a drop of oil on the breaker arm pivot.

GENERATOR

The Generator Bearings are prelubricated and require no attention.

STARTER

The oil hole cover on Commutator (Front) End slips to one side; 3 to 5 drops of medium engine oil is recommended every 1,000 miles, No. 36, Fig. 1. Be sure to slip cover back in place.

UNIVERSAL JOINTS—(Propeller Shaft)

Every 1,000 miles lubricate the propeller shaft universal joints with a hand gun and adaptor using S.A.E. 140 No. 6, Fig. 1. Use NLGI No. 1 in the slip joint, No. 7, Fig. 1.

UNIVERSAL JOINTS—(Front Axle Shaft)

Front axle shaft universal joints should be checked every 1,000 miles thru plug hole in rear of housing and add NLGI No. 1 lubricant to level of filler plug. Every 12,000 miles remove, clean, inspect and refill with ½ lb. No. 27, Fig. 1.

WHEEL BEARINGS

Wheel Bearings should be removed, thoroughly cleaned and repacked every 6,000 miles, No. 19, Fig. 1, using NLGI No. 1 lubricant

TRANSMISSION AND TRANSFER CASE

Transmission and Transfer cases are filled with S.A.E. 90 mineral oil lubricant at the factory, this being satisfactory for "year round" use except where extremely cold temperatures are experienced in which case use S.A.E. 80 or the oil should be diluted 10% to 20% with Kerosene. It should be checked each 1,000 miles when vehicle is lubricated and renewed each 6,000 miles. See Fig. 1, No. 28 and Fig. 29. To drain transmission remove skid plate. Lubricate transfer case shift lever shaft hydraulic fitting with NLGI No. 1 lubricant.

FRONT AND REAR AXLE

Hypoid gears require extreme pressure lubricant, therefore the lubricant manufacturers have developed a special lubricant, which is suitable for hypoid axles.

The level of the lubricant in these units should be checked every 1,000 miles. Do not mix different types of Hypoid Lubricants.

Seasonal changes of lubricant are not required except where extremely cold temperatures are experienced in which case use S.A.E. 80 or the oil should be diluted 10% to 20% with Kerosene. It is recommended that the housings be drained and refilled with 2½ pts. of S. A. E. 90 EP lubricant at least twice a year or every 6,000 miles. Use a light engine oil or flushing oil to clean out housings. See Fig. 1, No. 30.

Note—Do not use water, steam, kerosene, or gasoline for flushing.

AIR CLEANER

Each engine oil change or oftener, when vehicle is operated in sandy or dusty areas, remove, clean and refill oil cup to indicated oil level. See Lubrication Chart on Page 12. To clean element, see Page 50.

PERIODIC INSPECTION

OPERATION	Daily	Each 1000 Miles	Each 6000 Miles	12,000
Front Axle				
Check Wheel Alignment		X		
Inspect Tie Rod Ends for Wear		X		
Inspect Steering Arms for Tightness		X		
Inspect Steering Knuckle Bearings and Oil Seals for Looseness and Wear		X		
Check Axle Shaft Universal Joints for Wear			X	
(Make same Inspections as for Rear Axle)				
Rear Axle				
Check Axle Shaft Flange Bolts for Tightness		X		
Check Wheel Bearings for Looseness and Wear		X		
Inspect for Oil Leaks at Pinion Shaft Oil Seals		X		
Check Axle Housing Cover Bolts		X		
Inspect Axle Shaft Oil Seals for Grease Leak		X		
Check for End Play in Pinion Shaft		X		
Check Universal Joint Flanges for Looseness		X		
Body				
Check Body Bolts		X		
Brakes				
Inspect Fluid Supply in Master Cylinder		X		
Make Visual Inspection of Brake Lines and Hoses	X			
Test Service Brakes; adjust if necessary	X			
Remove Wheels; inspect brake lining			X	
Flush entire system with new fluid				X
Check Brake Pull Back Springs		X		
Test Hand Brake	X			
Clutch				
Check Free Pedal Travel; adjust when necessary	X			
Check Adjustment on Clutch Cable to determine when Driven Plate is required—(Last adjustment)		X		
Cooling				
Check Water in Radiator	X			
Test Anti-Freeze Solution (During Winter)	X			
Check Fan Belt for Tension	X			
Inspect Radiator Hoses and Connections	X			
Check Water Pump for Leaks	X			
Check Temperature Gauge	X			
Flush System Twice a year. Before and after using Anti-Freeze; test Thermostat; replace Radiator Hoses				X
Battery				
Check Gravity and add Distilled Water every 2 weeks				
Check Terminals		X		
Check and Tighten Ground Straps		X		
Check Hold Clamp Bolts		X		
Wiring				
Inspect all Connections		X		
Inspect for Chaffed or Broken Wires		X		
Inspect Retaining Clips and Grommets		X		

OPERATION	Daily	Each 1000 Miles	Each 6000 Miles	12,000
Starting Motor				
Clean Commutator with .00 Sand Paper			X	
Check Brushes for Wear and Tension			X	
Check Mounting Bolts for Tightness			X	
Overhaul				X
Clean Bendix Drive				X
Check Cable Connections			X	
Generator				
Clean Commutator with .00 Sand Paper			X	
Check all Terminals			X	
Check Brushes for Wear and Tension			X	
Overhaul				X
Check Voltage Regulator			X	
Lights and Switches				
Check Operation of Lights	X			
Horn				
Check for operation	X			
Distributor and Spark Plugs				
Clean and Adjust Distributor Points and Spark Plug Gaps			X	
Replace Plugs				X
Check and Test Condenser			X	
Overhaul Distributor				X
Check Timing			X	
Check Automatic Spark Advance			X	
Clean Distributor Cap—Terminal Towers			X	
Test Coil (Heat Test)				X
Engine				
Check Cylinder Head Nuts and Bolts			X	
Check Manifold Nuts and Gaskets		X		
Check Oil Pan Bolts; Check for Oil Leaks		X		
Check Compression			X	
Check Tappet Clearance			X	
Check Engine Mounting Bolts and Nuts			X	
Check Oil Pressure Gauge Reading	X			
Check Oil Level. (Add Necessary)	X			
Remove and Clean Oil Pan and Floating Intake				X
Change Oil Filter				X

Definite Periods for Major operations or overhauling cannot be predetermined. They are dependent upon service to which Engine has been subjected.

OPERATION	Daily	Each 1000 Miles	Each 6000 Miles	12,000
Fuel System				
Remove, clean and refill air cleaner		X		
(Vehicles operated in Dusty or Sandy regions, the air cleaner should be inspected and cleaned if necessary)	X			
Clean Fuel Pump Sediment Bowl and Strainer		X		
Clean Fuel Filter		X		
Tighten Carburetor Flange to Manifold Nuts		X		
Check Carburetor Adjustments		X		
Inspect all Fuel Lines and Connections for leaks		X		

PERIODIC INSPECTION—Continued

OPERATION	Daily	Each 1000 Miles	Each 6000 Miles	12,000
Fuel System				
Test Fuel Pump Pressure			X	
Overhaul Carburetor and adjust			X	
Drain Fuel Tank and Flush out Sediment				X
Lubrication				
Refer to Lubrication Chart	X	X	X	X
Springs				
Inspect Spring Clips to Axle for Tightness		X		
Inspect Spring Shackles and Bushings		X		
Check condition Front and Rear Springs		X		
Shock Absorbers				
Inspect Mounting Bushings, replace when necessary		X		
Inspect Mounting Brackets		X		
Check for Control (off car) adjust or replace				X
Steering System				
Check Steering System (Loss Motion)	X			
Inspect Steering Connecting Rod, Ball and Sockets		X		
Check Steering Post Bolts to frame		X		
Check Steering Pitman Arm Nut		X		
Transmission and Transfer Case				
Inspect for Oil Leaks		X		
Check Transmission to Bell Housing Bolts			X	
Check Oil Seals at Propeller Shafts		X		
Universal Joints				
Check Flange Nuts		X		
Inspect Joint Bearings for wear		X		
Wheels and Tires				
Check Tire Pressures	X			
Tighten Wheel Hub Bolt Nuts		X		
Remove Wheel Bearings, inspect, replace worn or chipped cups or cone repack and adjust			X	
Check Tire Wear, check toe-in, caster and camber		X		

Inspection

The importance of regular inspection cannot be over-emphasized. Making adjustments, tightening bolts, nuts and wiring connections when needed, will go far towards avoiding trouble and delay on the road and uphold the high standards of reliability and performance built into the vehicle by the Manufacturer.

After maneuvers involving operations in swamps and streams inspect for water and sludge in engine, transmission, transfer case, front and rear axles, wheel bearings and front universal joints.

ENGINE

(NEVER RUN ENGINE IN CLOSED GARAGE)

Due to the presence of carbon monoxide (a poisonous gas in the exhaust of the engine) never run the engine for any length of time while this vehicle is in a small closed garage. Opening the doors and windows lessens the danger considerably, but it is safest, if adjustments are being made that require operation of engine, to run the vehicle out of doors.

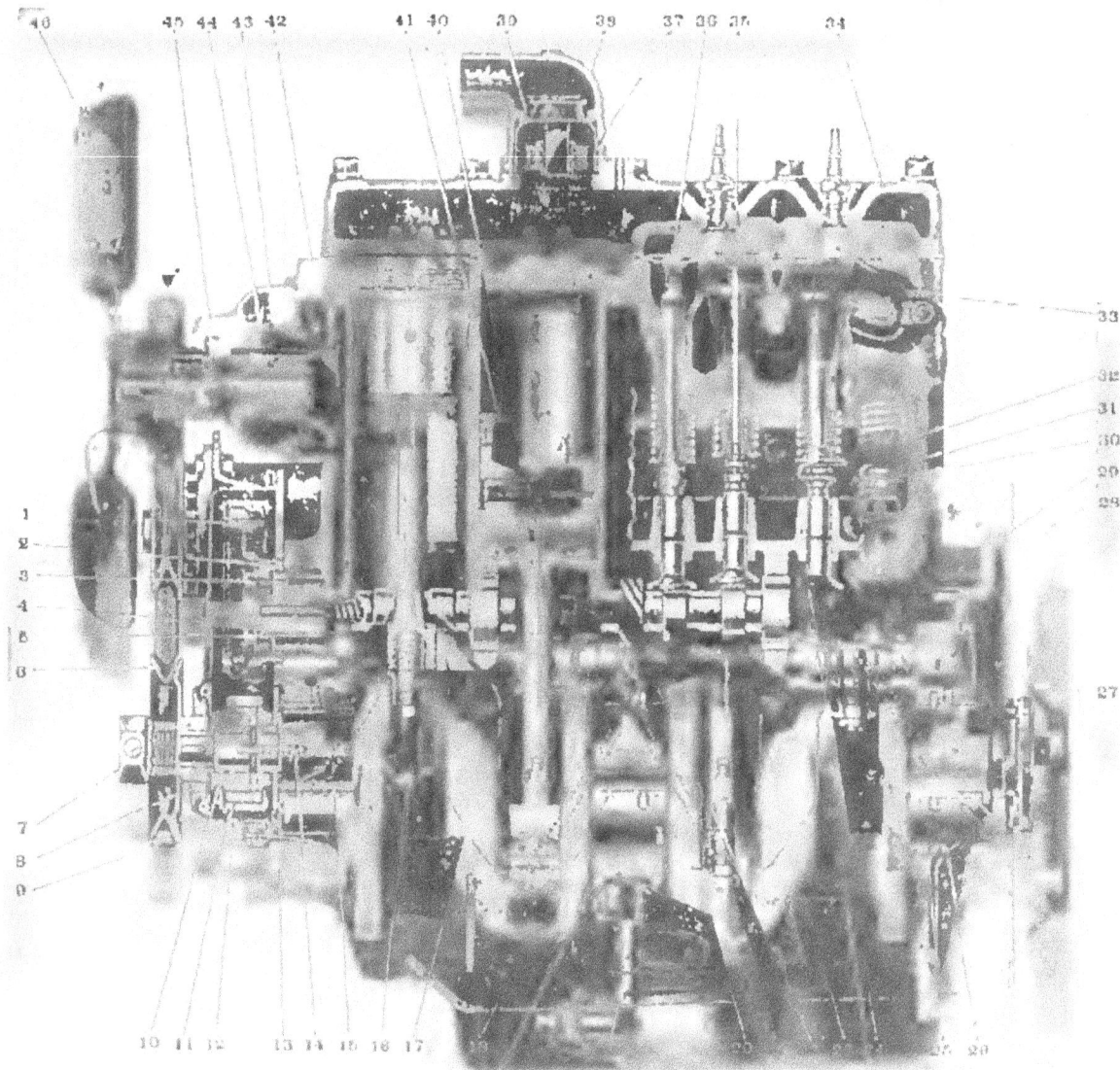

FIG. 1—SIDE SECTIONAL OF ENGINE

The engine is a Four-Cylinder L head type unit equipped with a counter-balanced crankshaft.

At end of this section will be found the Engine specifications. When adjustments are necessary we recommend that reference be made to these specifications for proper running tolerances and clearances of all component parts. On Page 34 headed "Engine Troubles and Causes" are listed many reasons for engine failure or poor performance. For correction of these difficulties you will find the procedure to follow under separate headings in this section.

FIG. 1—SIDE SECTIONAL VIEW OF ENGINE

No.	Willys Part No.	Ford Part No.	Name
1	638458	GPW-6256	Camshaft Sprocket
2	375900	GPW-6345	Camshaft Thrust Washer
3	639051	GPW-6262	Camshaft Front Bushing
4	375907	GPW-6243	Camshaft Thrust Plunger
5			Fuel Pump Eccentric
6	638113	GPW-6312	Fan and Generator Drive Pulley
7	387633	GPW-6319	Starting Crank Nut Assembly
8	637098	GPW-6700	Crankshaft Packing—Front End
9	A-1495	GPW-8620	Fan and Generator Drive Belt
10	638459	GPW-6306	Crankshaft Sprocket
11	638457	GPW-6260	Camshaft Drive Chain
12	A-1190	GPW-6016	Chain Cover Assembly
13	634796	GPW-6308	Crankshaft Thrust Washer
14			Crankshaft Oil Passages
15	637008	GPW-6338-A	Crankshaft Bearing Front—Lower
16	52825	356028-S	Connecting Rod Bolt Nut Lock
17	639859		Connecting Rod Assembly No. 2
18	638731	GPW-6341-A	Crankshaft Bearing Center—Lower
19	630396	GPW-6618	Oil Float Assembly
20	630397	GPW-6617	Oil Float Support
21	637020		Connecting Rod Cap Bolt
22			Oil Pump and Distributor Drive Gear
23	638121	GPW-6303-A1	Crankshaft
24	637047	GPW-6500	Valve Tappet
25	638733	GPW-6337-A	Crankshaft Bearing Rear—Lower
26	630294	GPW-6326	Crankshaft Bearing Rear Drain Pipe
27	637237	GPW-6702	Crankshaft Packing—Rear End
28	635394	GPW-6384	Flywheel Ring Gear
29	637065	GPW-6250	Camshaft
30	637704	GPW-6550	Valve Tappet Clearance Spring
31	637048	GPW-6549	Valve Tappet Adjusting Screw
32	638636	GPW-6513	Valve Spring
33	A-912		Exhaust Manifold Assembly
34	A-1534	GPW-6050	Cylinder Head
35	637182	GPW-6507	Inlet Valve
36	637183	GPW-6505	Exhaust Valve
37	639651	GPW-8678	Thermostat Retainer
38	A-1192	GPW-8250	Water Outlet Elbow
39	637646	GPW-8575	Thermostat Assembly
40	635961	GPW-6135-A	Piston Pin
41	635957	GPW-6110-A	Piston
42	639993	GPW-8512	Water Pump Impeller
43	639563	GPW-8524	Water Pump Seal Assembly
44	639994	GPW-8557	Water Pump Seal Washer
45	638207	GPW-8530	Water Pump Bearing and Shaft Assembly
46	A-447	GPW-8600	Fan Assembly

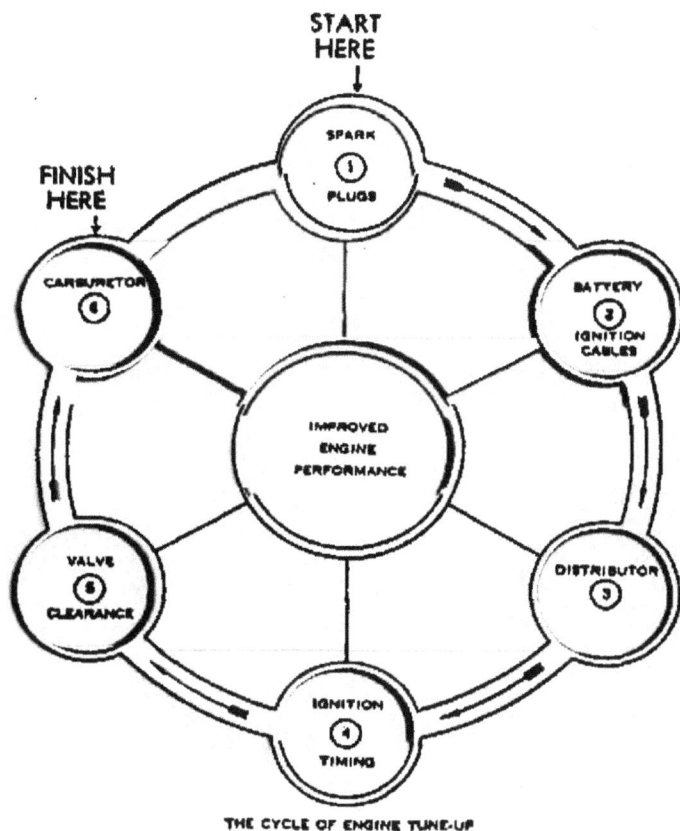

FIG. 2

THE CYCLE OF ENGINE TUNE-UP

Engine Tune-Up

For best performance and dependability, the engine should have a periodic tune-up every 6,000 miles, see Fig. 2. The following procedure is recommended when performing this operation:

1. Remove spark plugs and clean. Adjust the Electrodes to .030" gap.
2. Check Battery Terminals, ground cable and ground straps on left side of engine at front engine support and cylinder head for clean and tight connections.
3. Remove distributor cap and inspect points.
4. Check ignition timing.
5. Check valve tappet clearance-set .014" cold.
6. Set carburetor float level, accelerator pump travel, and metering rod as covered under heading "Carburetor".
7. Start engine and allow to run until thoroughly warmed up, then set carburetor idle screw so the engine will idle at 600 R.P.M. (vehicle speed approximately 8 miles per hour).
8. Adjust low speed idling screw so that engine will idle smoothly.

Carburetor

Complete information regarding dismantling, cleaning and adjusting will be found in the Fuel section, under heading "Carburetor".

Distributor

For complete information regarding cleaning, adjusting and setting ignition timing refer to Electrical section under heading "Distributor".

Grinding Valves

Lack of power in an engine is sometimes caused by poor seating of the valves in the valve seats which allows the gases in the compression chamber to escape into the intake or exhaust manifold.

Through the use of a cylinder compression gauge one can readily determine which valves are not properly seating.

Compression gauge readings should all be within 10 pounds of each other and not less than 70 lbs.

If no gauge is available, remove all spark plugs, hand crank engine and have mechanic place thumb over one spark plug hole at a time. When no compression is experienced that particular cylinder is at fault.

The valve grinding operation from the standpoint of engine power and performance is very important. Extreme care should be used whenever valves are ground to maintain factory limits and clearances, as only by maintaining these can one expect to get good engine performance.

When it is necessary to perform a valve job, it will be best to follow the procedure as outlined in the next few paragraphs.

1. Drain radiator by opening drain cock at the bottom left corner of the radiator.
2. Remove oil filter and bracket by removing the nuts on the cylinder studs and lay filter on generator.
3. Remove fuel line from fuel pump to carburetor.
4. Remove carburetor air cleaner and accelerator rod.
5. Remove choke and throttle control wires.

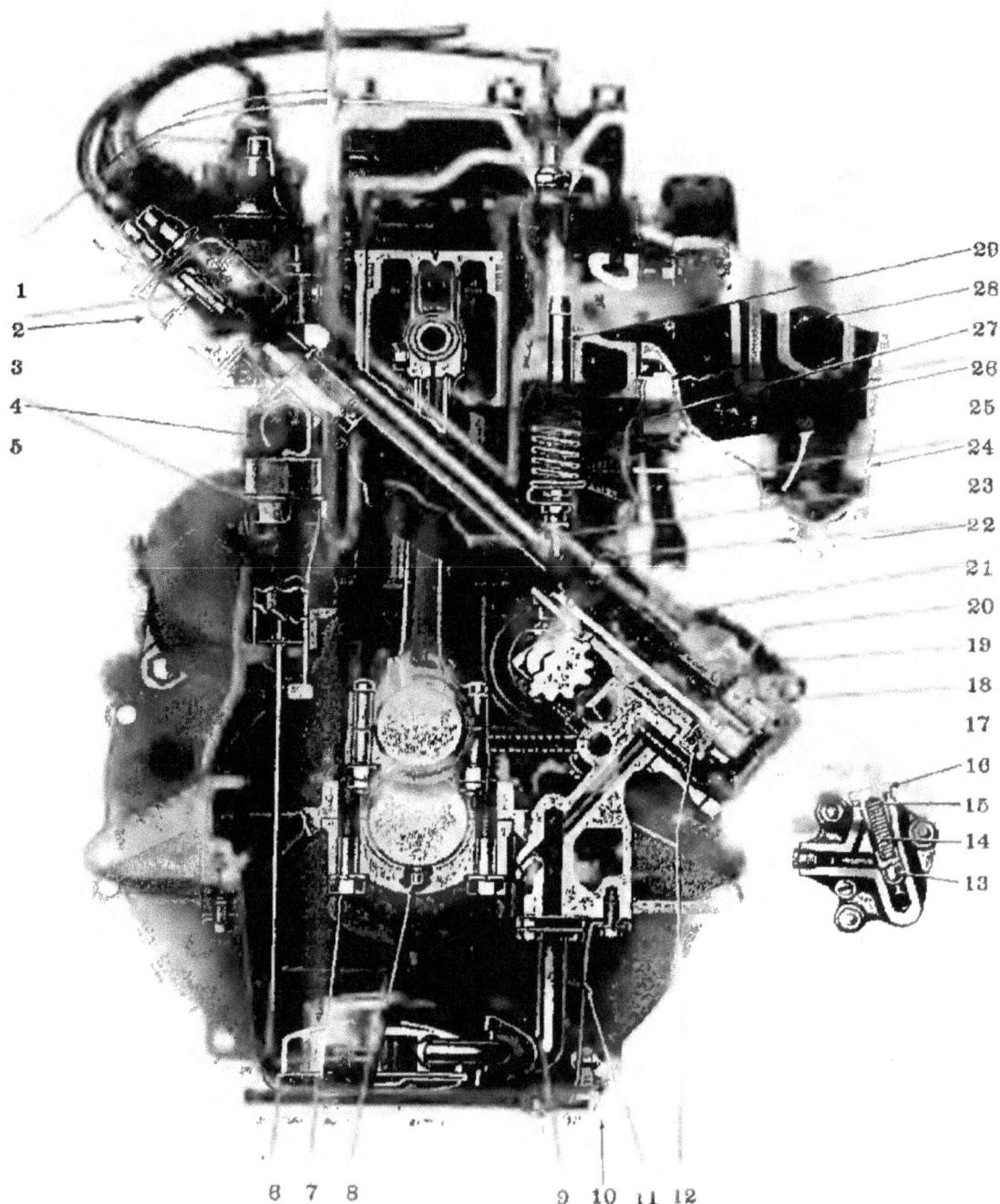

FIG. 3—FRONT SECTIONAL OF ENGINE

No.	Willys Part No.	Ford Part No.	Name	No.	Willys Part No.	Ford Part No.	Name
1	A-1527	GPW-12000	Ignition Coil	15	630389	GPW-6628	Oil Relief Plunger Spring Shims
2	A-1244	GPW-12100	Ignition Distributor	16	630390	GPW-6644	Oil Relief Plunger Spring Retainer
3	107128	B-10141	Distributor Oiler	17	343306	GPW-6614	Oil Pump Pinion
4	A-5168	GPW-6766-B	Oil Filler Cap and Level Indicator	18	637636	GPW-6600	Oil Pump Assembly
5	A-5165	GPW-6763-B	Oil Filler Tube	19	636600	GPW-6673	Oil Pump Rotor Disc
6	630396	GPW-6615	Oil Float Assembly	20	630386	GPW-6609	Oil Pump Shaft
7	381519	GPW-6345	Crankshaft Bearing Cap to Crankcase Screw	21	630394	GPW-6630	Oil Pump Body to Cylinder Block Gasket
8	635377	GPW-6369	Crankshaft Bearing Dowel	22	637615	GPW-12083	Distributor Shaft Friction Spring
9	630397	GPW-6617	Oil Float Support	23	A-1061	GPW-6758	Crankcase Ventilator Assembly
10	639979	GPW-6727	Oil Pan Drain Plug	24	A-912		Exhaust Manifold Assembly
11	A-1167	GPW-6675	Oil Pan Assembly	25	630298	GPW-6762	Crankcase Ventilator Baffle
12	636599	GPW-6608	Oil Pump Shaft and Rotor Assembly	26	636439	GPW-9460	Heat Control Valve
13	630518	GPW-6663	Oil Relief Plunger	27	636554	GPW-6519	Valve Spring Cover Assembly
14	356155	GPW-6654	Oil Relief Plunger Spring	28	A-1166		Intake Manifold Assembly
				29	375811	GPW-6510	Exhaust Valve Guide

6. Remove nuts holding carburetor to manifold and remove carburetor.

7. Remove bolt and nuts holding exhaust pipe to manifold.

8. Remove manifold stud nuts and manifolds.

9. Remove the upper radiator hose. Remove all spark plugs by using the socket wrench furnished in the tool kit. Remove the cylinder head cap screws, stud nuts and the temperature gauge bulb, then lift head from engine block. Removal is made easy by using lifting hooks screwed in No. 1 and 4 spark plug holes. CAUTION—Do not use screw driver or any other sharp instrument to drive in between the cylinder head and the block to break the head loose from the gasket.

10. Remove the valve cover plate screws and the valve cover. Care should be taken when removing the valve cover breather tube not to lose the copper gasket on each screw as well as the screen and gasket. With a piece of cloth or cotton waste cover the three holes in the valve chambers to prevent the valve keys dropping into crankcase upon removal.

11. Remove valve tappet clearance springs by placing screw driver on top edge and snapping out. With valve spring compressor inserted between valve tappet and spring retainer raise springs on those valves which are in closed position and remove valve locks. Turn crankshaft with crank or by fan belt until those valves which are open become closed and repeat the operation.

12. Remove valves and place them in a valve carrying board, so that they can be identified as to cylinders from which they were removed. Remove valve springs. The valve springs should be tested for pressure which should show 116 lbs. when valves are open (Length 1¾") or 50 lbs. pressure when closed (Length 2⁷⁄₆₄"). The free length of the valve spring is 2½" inches. Any springs which are distorted or do not fall within these specifications should be replaced with new springs.

13. Clean carbon from cylinder head, top of pistons, valve seats and cylinder block, clean valve guides with guide brush. Clean valves on a wire wheel brush making sure that all carbon is removed from the top and bottom of the heads, as well as the gum which might have accumulated on the stems. The valve heads should then be refaced at an angle of 45°. If valve seats in block show signs of excessive pitting it is advisable to reface the seats and check with dial gauge—Fig. 4. Then by hand, touch up the valves to the valve seats with fine valve grinding compound.

The clearance between the intake valve stem and the valve guide is .0015" to .00325", the exhaust valve stem clearance to guide is .002" to .00375". Excessive clearance between the valve stem and the valve guide will cause improperly seating and burned valves. If there is too much clearance between the inlet valve stem and the valve guide, on the suction stroke there will be a tendency to draw oil vapors up the guide into the combustion chamber causing excessive oil consumption, fouled spark plugs and poor low speed performance.

To check the clearance of the valve stem to the valve guide, take new valve and place in each valve guide and feel the clearance by moving the valve stem back and forth. If this tolerance is excessive it will be necessary to replace the valve guide, otherwise the valve stem is worn.

FIG. 4—GAUGING VALVE SEATS

FIG. 5—PULLING VALVE GUIDE

Removing and Replacing of Valve Guide

When removing the valve guides use a valve guide puller such as shown in Fig. 5 to prevent damage to cylinder block. If a regular puller is not available, a suitable tool can be made from a 2" pipe, 6" long and a ⅜" bolt 10" to 12" long with a long threaded end, a small hexagon nut which will pass through the hole in the cylinder block and a 2" washer with a ⅜" hole in it.

The valve guides are installed with a replacer or a driver as shown in Fig. 6. Taking a piece of half inch round stock 6" long and turning down one end to ⅜" diameter 2" long will make a suitable driver.

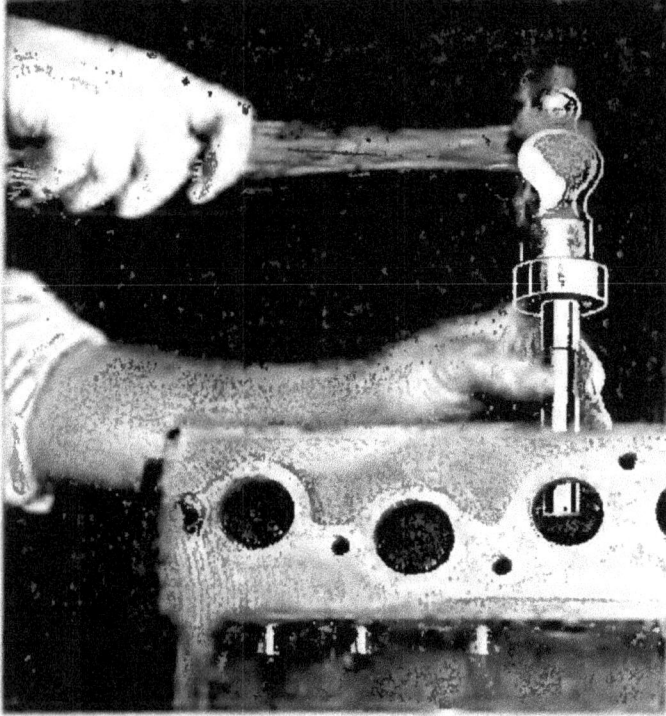

FIG. 6—INSTALLING VALVE GUIDE

The exhaust valve guide is installed in the cylinder block so that there will be a distance of 1" from the top of the guide to the top of the block. The intake valve guide is set at 1 5/16" from top of valve guide to the top of block. Fig. 7.

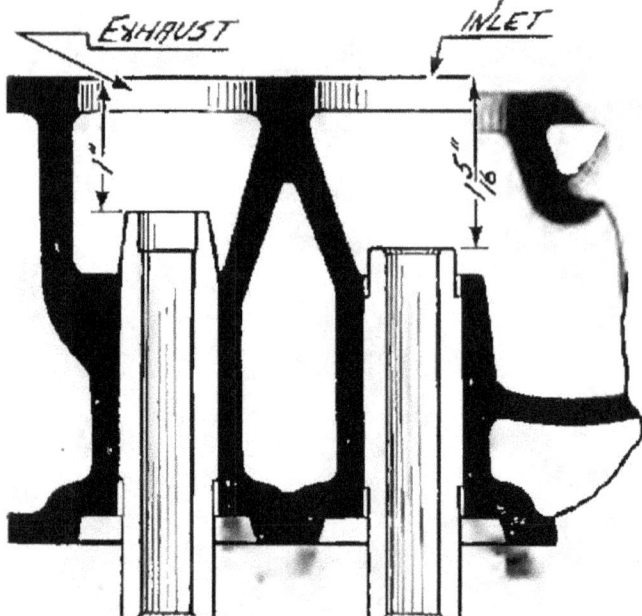

FIG. 7—POSITION OF VALVE GUIDES

The valve tappet clearance in the guide should be .0005" to .002". It is advisable to check the clearance of the valve tappet by moving it back and forth in the guide. If the clearance seems to be excessive it might be necessary to install a new valve tappet. This operation is covered in this section under "Camshaft and Valve Tappet."

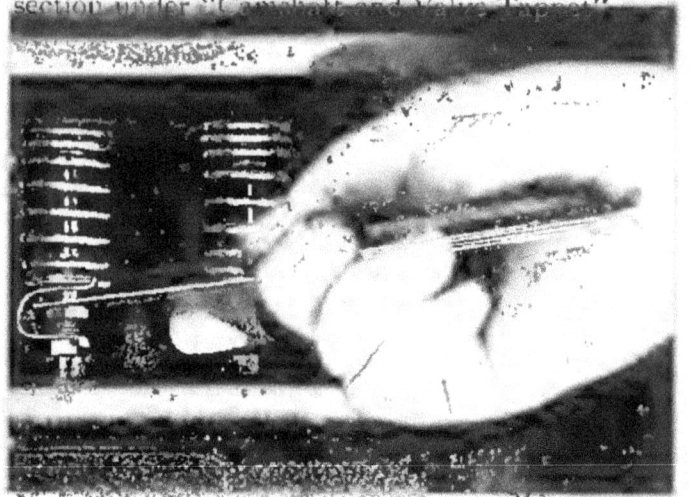

FIG. 8—VALVE TAPPETS AND SPRINGS

When assembling valve springs and retainers in engine make sure that the closed coils are up against cylinder block. See Fig. 8. Then make installation of the valves in their respective positions as they were disassembled. Through the use of a valve spring compressor raise valve springs on those valves which are in the closed position, and with valve key inserting tool, insert the valve spring locks. If no tool is available, hold keys in place by sticking them to valve stem with grease.

Adjust the valve tappet to valve stem clearance to .011". Fig. 8. Remove cloth or waste from valve chamber.

Clean top of block and pistons of all foreign matter and install cylinder head gasket. Clean carbon from cylinder head and wipe off all foreign matter then install over studs on cylinder block.

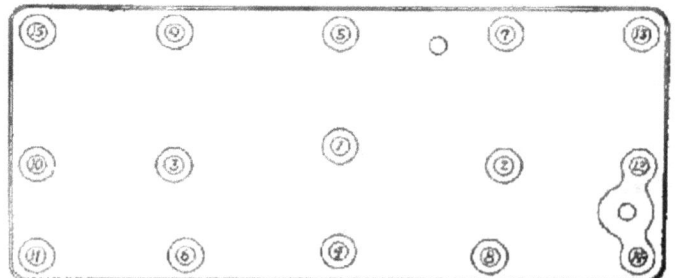

FIG. 9—CYLINDER HEAD TIGHTENING

Install oil filter and air cleaner bracket and tube assembly. Install cylinder head cap screws and nuts bringing them down finger tight, then with a tension wrench tighten cylinder head screws and nuts in sequence as shown on Fig. 9, tightening screws to 65 to 75 foot pounds or 780 to 920 inch pounds and the nuts to 60 to 65 foot pounds or 720 to 780 inch pounds.

Clean and adjust spark plugs, setting the electrode gaps at .030", Fig. 10. Install spark plugs in cylinder head to prevent any foreign matter from entering the combustion chamber during the remaining operations.

Install manifold with new gaskets. Install manifold clamp washers with convex surface toward manifold. Install manifold nuts drawing them up tight. Install exhaust pipe to manifold with new gasket.

Overhaul and recondition carburetor as per instructions given under the heading "Carburetor." Install carburetor to manifold and attach controls.

Recondition distributor and set ignition timing in accordance with instructions given under "Distributor".

NOTE—Make sure when installing distributor assembly in crankcase that it fits down in the crankcase properly.

Install upper radiator hose, and all line connections and fill radiator with water. Start engine and allow it to idle for a period of five or ten minutes, then recheck tappet clearance.

If necessary, install new valve cover plate gasket (shellac to cover). Install cover plate to engine block. Clean valve chamber ventilator tube and screen and reinstall with gaskets.

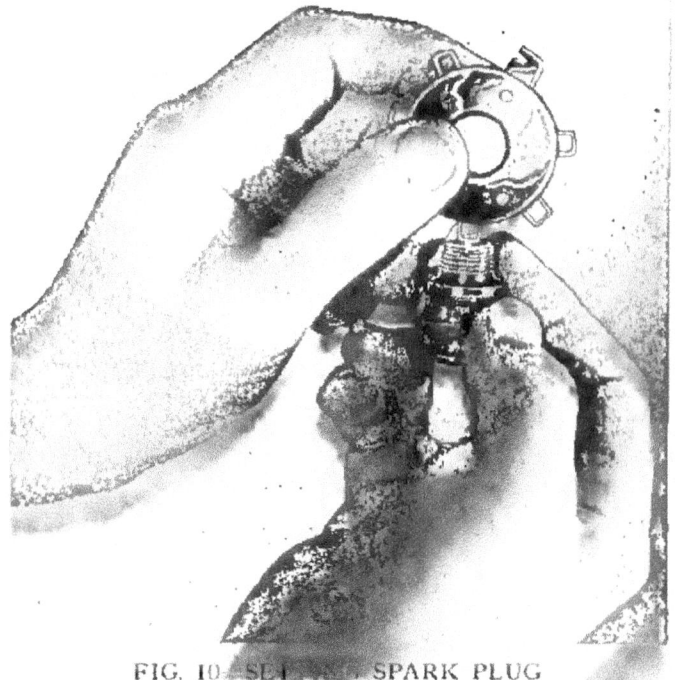

FIG. 10—SETTING SPARK PLUG

FIG. 11— VALVE MECHANISM

No.	Willys Part No.	Ford Part No.	Name
1	637183	GPW-6505	Exhaust Valve
2	638636	GPW-6513	Valve Spring
3	637044	GPW-6514	Valve Spring Retainer Lower
4	375904	GPW-6546	Valve Spring Retainer Lower Lock
5	637704	GPW-6550	Valve Tappet Clearance Spring
6	637048	GPW-6549	Valve Tappet Adjusting Screw
7	375910	355571-S	Valve Tappet Adjusting Screw Lock Nut
8	637047	GPW-6500	Valve Tappet
9	639051	GPW-6262	Camshaft Bushing—Front
10	637065	GPW-6250	Camshaft
11	375900	GPW-6245	Camshaft Thrust Washer
12	638457	GPW-6260	Camshaft Drive Chain
13	638458	GPW-6256	Camshaft Sprocket
14	638459	GPW-6306	Crankshaft Sprocket
15	316932	GPW-6269	Camshaft Sprocket Cap Screw Lockwasher
16	634850	355499-S	Camshaft Sprocket Cap Screw
17	375903	GPW-6244	Camshaft Thrust Plunger Spring
18	375907	GPW-6243	Camshaft Thrust Plunger

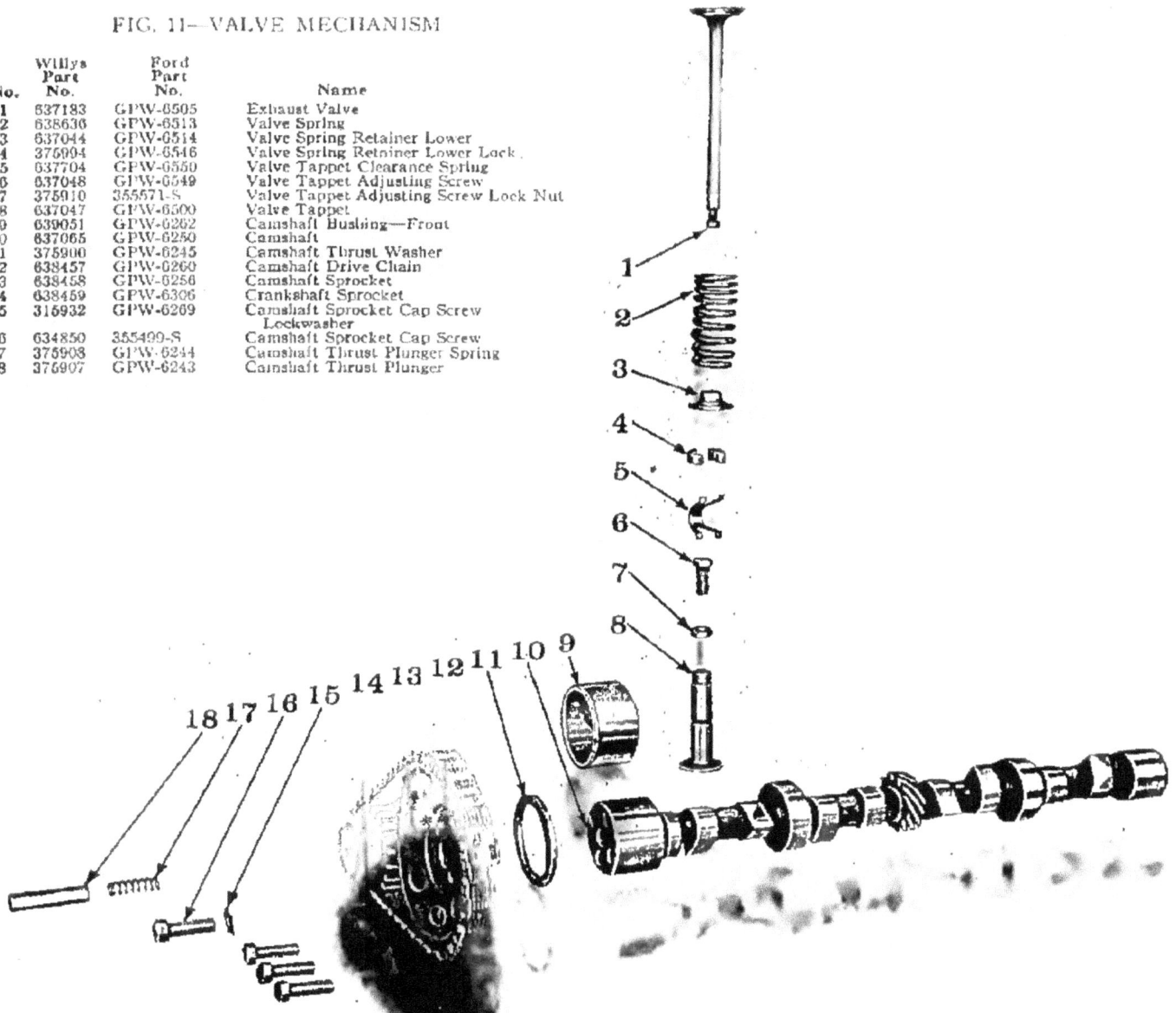

Camshaft and Valve Tappets

The alloy cast steel camshaft Fig. 11 rotates on four bearings which are lubricated under oil pressure through drilled passages in the crankcase. The front bearing carries the thrust and is a steel back babbitt-lined shell. This bearing is staked in place to prevent rotation and endwise movement. See Fig. 12.

The valve tappets are lubricated through oil troughs cast in crankcase and drilled passages to valve tappet guides. The oil troughs are filled from oil spray holes at connecting rod bearing ends. A groove cut in center of valve tappet shank carries the oil up and down in guides.

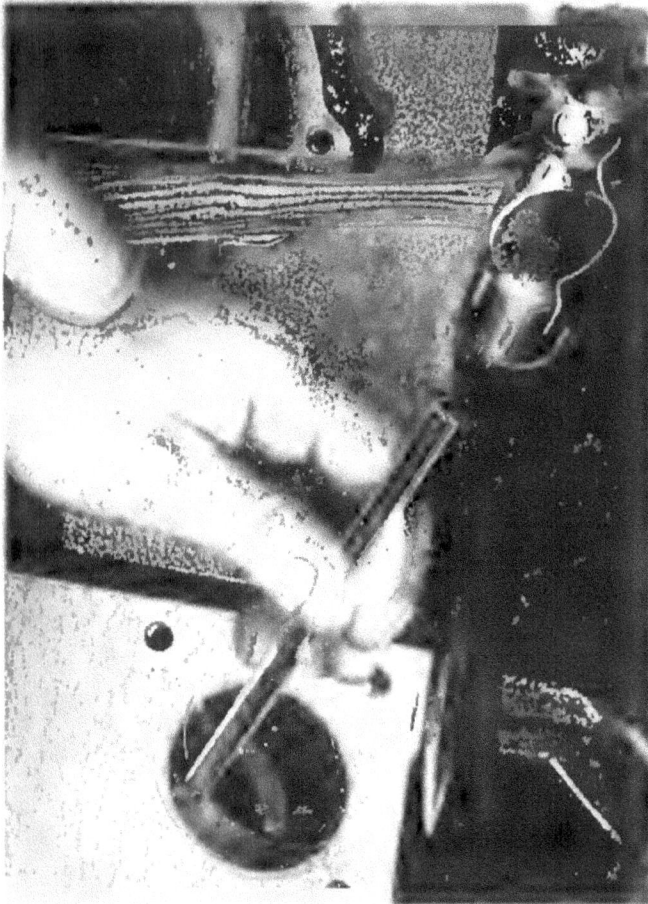

FIG. 12—STAKING CAMSHAFT BEARING

Removal of Camshaft or Valve Tappet

Drain water from radiator, remove radiator and grille, cylinder head, manifold, valves and valve springs. Follow instructions given under sub-heading "Grinding Valves" Page 18.

Remove oil pump and fuel pump assemblies.

Remove oil pan, crankshaft pulley, fan belt and fan assembly.

Remove nuts holding front engine supports to rubber insulators.

Remove timing chain cover, camshaft sprocket screws and timing chain.

Tie all valve tappets up with a string wrapped around heads of screws and attach to manifold studs.

Place jack with block under crankcase and raise front end of engine until camshaft will clear front

frame cross member. Remove camshaft and valve tappets.

Carefully inspect camshaft for scores, roughness of cams and bearings. Examine valve tappet faces where they contact cams and replace if found to be scored, rough or cracked. Check clearance of tappets to guides, renewing those which have worn excessively. Oversize available .004".

Replacing Camshaft or Valve Tappets

Install valve tappets and tie up in place with string. Install camshaft. Install camshaft thrust washer.

To set the valve timing, see instructions given under heading "Valve Timing" Page 24.

For installation of oil pump see section under heading "Oil Pump" Page 29.

Install the plunger and spring in the front end of camshaft with round end out. Inspect pin in timing chain cover to see that it stands perpendicular to the cover face. Put a light smear of cup grease on end of pin and on the end of plunger, then assemble cover to the engine.

Balance of the assembly is the same operations used in removal of camshaft only in the reverse order.

NOTE—On earlier engines, when replacing oil pan, the four short cap screws are used across front end.

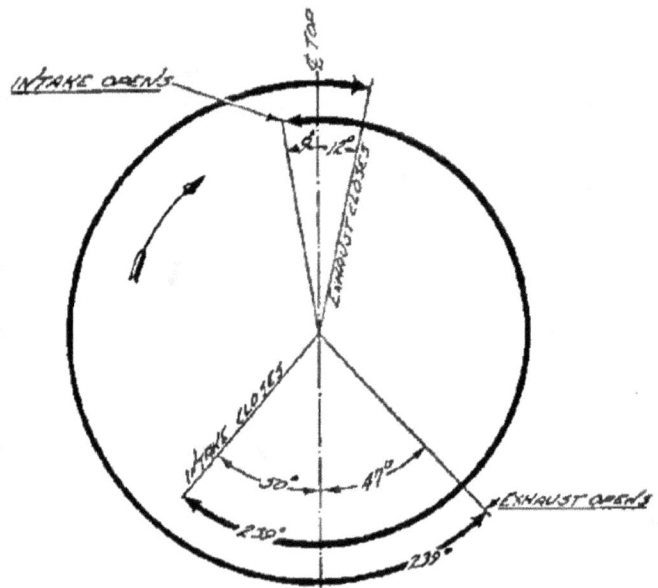

FIG. 13—VALVE TIMING

TIMING CHAIN AND SPROCKETS

The silent type timing chain is non-adjustable. The lubrication is positive through drilled passages in the crankshaft and sprocket from the front main bearing and oil filter return. These should be checked whenever the chain or sprockets are replaced.

To replace timing chain, it is necessary to remove fan blades, radiator, fan belt, crankshaft pulley and timing case cover. Remove screws holding camshaft sprocket to camshaft and remove chain.

When chain has been removed, for any purpose, it will be necessary to set the valve timing when chain is replaced.

VALVE TIMING

To set the valve timing turn the crankshaft so No. 1 and No. 4 pistons are at top dead center.

Place camshaft sprocket on camshaft, turn camshaft so punch mark faces punch mark on crankshaft sprocket.

Remove sprocket and place the timing chain on, then place timing chain over crankshaft sprocket, changing position of camshaft sprocket within chain until the cap screw holes in sprocket and camshaft are in line.

Timing is correct when a line drawn between sprocket centers cuts through timing marks on both sprockets. See Fig. 14.

Inlet opens 9° before top center measured on flywheel or .039″ piston travel from top center.

To check valve timing, Fig. 13 adjust inlet valve tappet No. 1 cylinder to .020″. Rotate crankshaft clockwise until piston in No. 1 cylinder is ready for the intake stroke, at which time the tappet should just be tight against end of valve stem and mark on flywheel "I O" is in center of timing hole in flywheel housing, see Fig. 15.

TIMING CHAIN COVER AND SEAL

The timing chain cover is a pressed steel stamping heavily ribbed for strength.

The stationary pin in cover is so located as to bear against the plunger in the end of camshaft which controls the end play of camshaft.

The crankshaft oil seal is woven asbestos impregnated with graphite and oil. When necessary to install new oil seal, the steel retainer should also be renewed.

FIG. 14—TIMING SPROCKETS

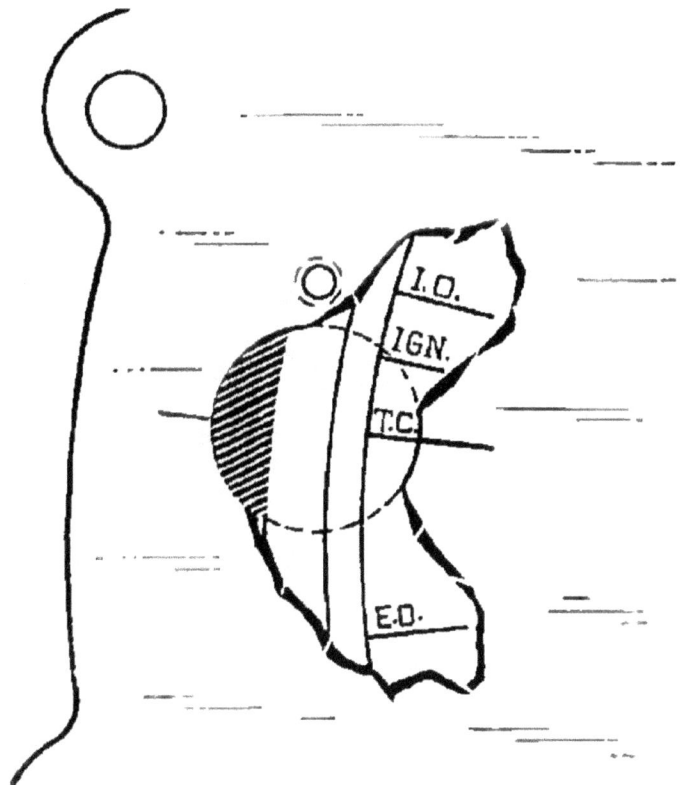

FIG. 15—TIMING MARKS (FLYWHEEL)

Crankshaft

The crankshaft Fig. 16 is of drop forged steel, grain direction following shape of crankshaft with four integral counterweights. This short shaft is rugged and reduces torsional vibration, weight 41½ lbs. After machining, the crankshaft is balanced statically and dynamically, and then dynamically balanced with clutch and flywheel as an assembly. It rotates on three steel back babbitt lined bearings, the front bearing taking the thrust. The packing at the rear bearing prevents oil getting into clutch and loss when parked on steep inclines.

The main bearing journal diameter and length dimensions are as follows:

Front—2.3340″—1.920″
Center—2.3340″—1¹³⁄₁₆″
Rear—2.3340″—1¾″

The steel back babbitt lined bearings are made to size and are interchangeable without line reaming. The running tolerance of the bearing is established at .001″. No adjustment is provided on the main bearings. Should they require attention they should be replaced to maintain proper control of oil. Main bearing cap screw torque wrench reading 65-70 ft. lbs.

The end play of the crankshaft is .004″ to .006″ and adjusted by shims between the crankshaft sprocket thrust washer and end of Main Bearing, Fig. 17.

To adjust end play the crankshaft sprocket must be removed with gear puller, Fig. 18.

Whenever it is necessary to remove the crankshaft or install new crankshaft bearings, the engine has to be removed from the frame. See Note under "Flywheel" Page 32.

Undersize main bearings are available in .010″; .020″ and .030″.

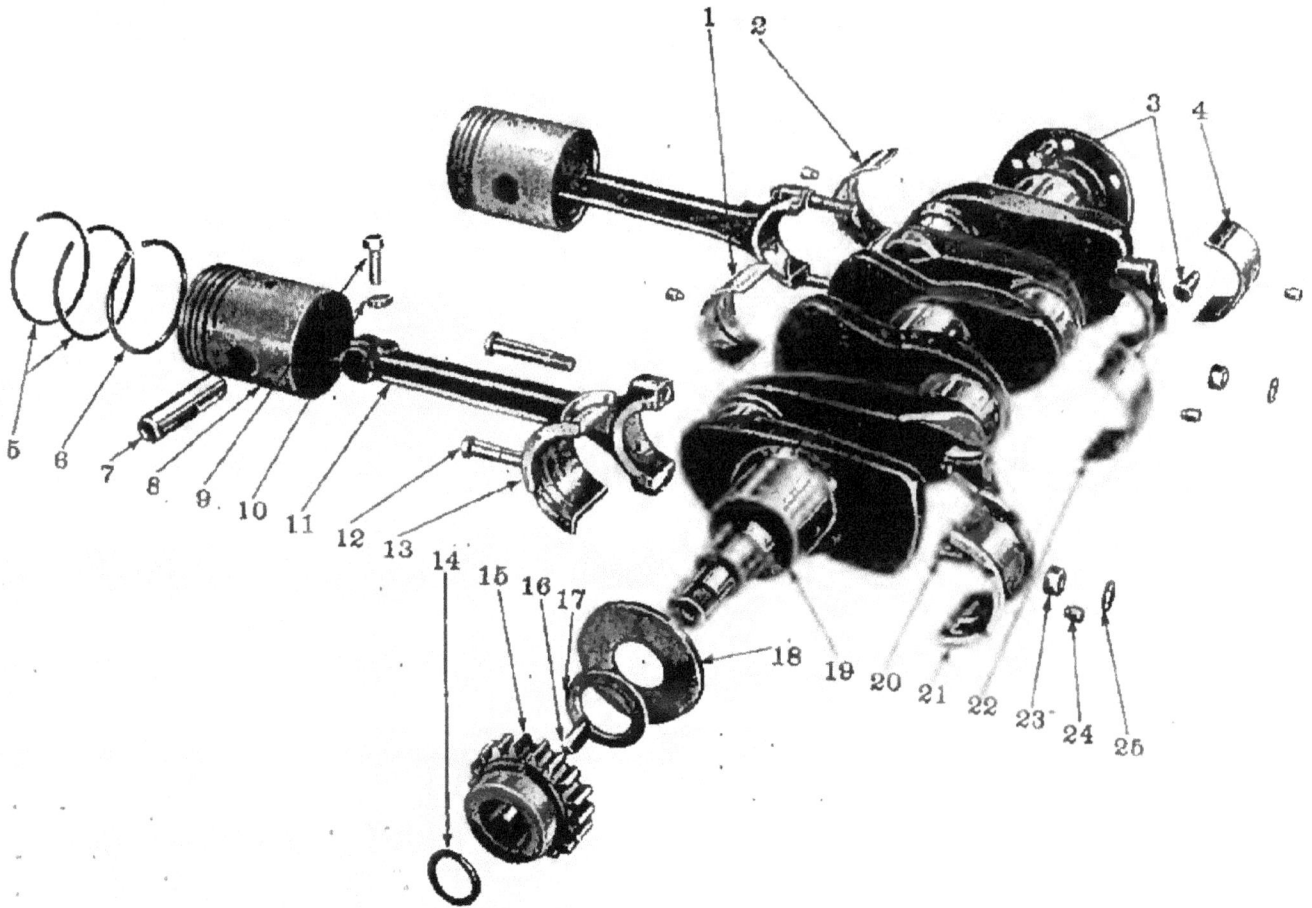

FIG. 16—CRANKSHAFT AND PISTONS

No.	Willys Part No.	Ford Part No.	Name
1	638730	GPW-6339-A	Crankshaft Bearing Center—Upper
2	638732	GPW-6331-A	Crankshaft Bearing Rear—Upper
3	632156	GPW-6387	Flywheel Crankshaft Dowel
4	638733	GPW-6337-A	Crankshaft Bearing Rear—Lower
5	116562	GPW-6155-A	Piston Ring—Compression (Lower)
	639864	GPW-6150	Piston Ring—Compression (Upper)
6	116566	GPW-6156-A	Piston Ring—Oil Regulating
7	638961	GPW-6135-A	Piston Pin
8	638957	GPW-6110-A	Piston—Grade "D"
9	632157	355497-S	Piston Pin Lock Screw
10	5010	34807-S	Piston Pin Lock Screw Lockwasher
11	639858	GPW-6200	Connecting Rod Assembly (No. 1 and No. 3 Cylinders)
12	637020		Connecting Rod Cap Bolt
13	637007	GPW-6333-A	Crankshaft Bearing Front—Upper
14	334103	GPW-6353	Crankshaft Oil Slinger Gasket
15	638459	GPW-6306	Crankshaft Sprocket
16	50917	74182-S	Crankshaft Sprocket Key
17	630727	GPW-6342-A	Crankshaft Sprocket Spacer
18	634796	GPW-6308	Crankshaft Thrust Washer
19	638121	GPW-6303-A	Crankshaft
20	639862	GPW-6211-A	Connecting Rod Bearing (Upper and Lower)
21	637008	GPW-6338-A	Crankshaft Bearing Front—Lower
22	638731	GPW-6341-A	Crankshaft Bearing Center—Lower
23	636962	356021-S	Connecting Rod Cap Bolt Nut
24	635377	GPW-6369	Crankshaft Bearing Dowel
25	52825	356023-S	Connecting Rod Cap Bolt Nut Lock

FIG. 17—CRANKSHAFT END PLAY

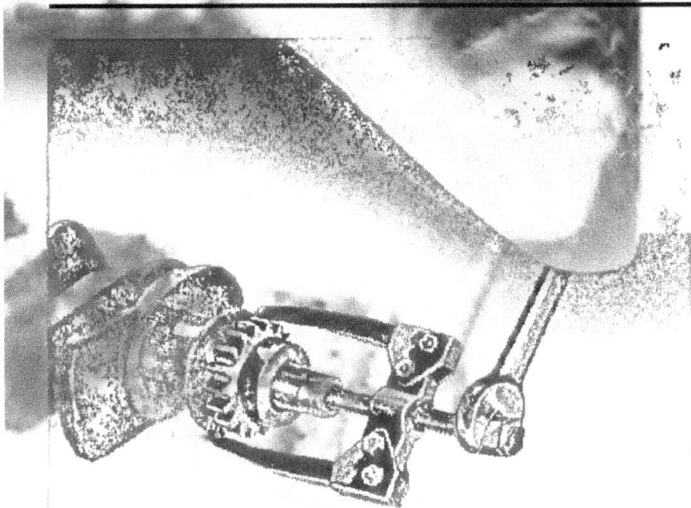

FIG. 18—CRANKSHAFT SPROCKET

CRANKSHAFT REAR BEARING SEAL

The rear main bearing is sealed by a wick type packing, installed in the groove machined in the crankcase, and rear main bearing cap, Fig. 19.

To install a new seal at the rear main bearing cap, insert the packing in the groove with the fingers. Then using a round piece of wood or steel, roll the packing into the groove. When rolling the packing, start at one end and roll the packing to the center of the groove. Then starting from the other end again roll towards the center. By following the above procedure you are sure that the wick is firmly pressed into the bottom of the groove.

The small portion of the packing which protrudes from the groove at each end should be cut flush with the surface of the bearing cap. To prevent the possibility of pulling the packing out of the groove while cutting off the ends it is recommended that a round block of wood, the same diameter as the crankshaft be used to hold the packing firmly in position while the ends are being cut off.

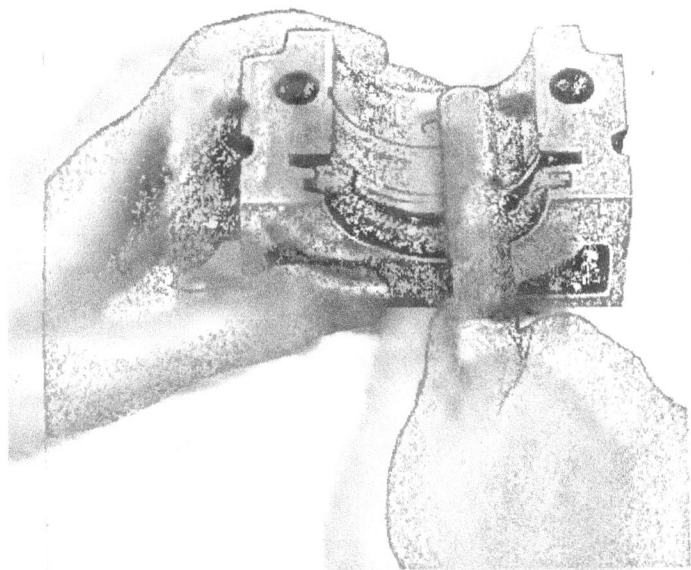

FIG. 19—REAR MAIN BEARING

Should it be necessary to install a new seal in the crankcase, it will require the removal of the engine from the frame and the removal of the crankshaft.

The same procedure should be followed when installing a crankcase seal as when installing a seal in the bearing cap.

When installing rear main bearing cap to case a little sealer should be put on the faces of cap where it fits against the case. The rubber seal packing that goes between the main bearing cap and the case is cut to a given length and will protrude down from the case approximately ¼". When the oil pan is installed it will force this seal tightly into the holes and prevent any oil from leaking from the engine into the clutch housing. See Fig. 20.

If new crankshaft bearings are installed care should be taken to see that the drilled passages line up with drilled passages in the crankcase, and that the bearings set snugly over the dowel pins.

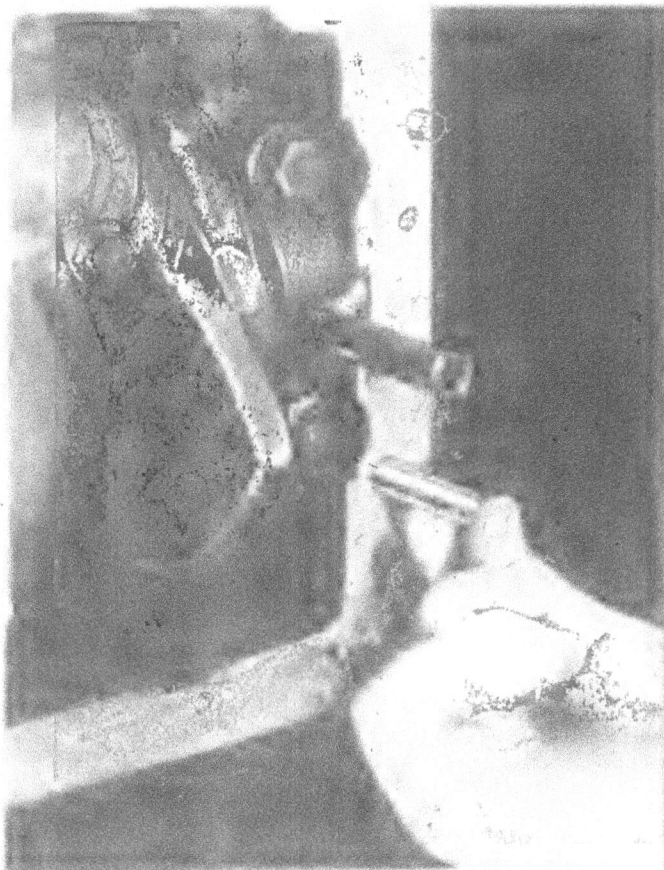

FIG. 20—REAR BEARING CAP SEAL

CONNECTING ROD

The connecting rods are drop forged and are of unusual length, measuring $9\frac{3}{16}$" from center to center. The babbitt bearings are of the replaceable type, steel-backed, babbitt lined, precision cut to size and no fitting is required.

When installing the upper half of rod bearing be sure that the oil spray hole lines up with the spray hole in the connecting rod. Undersize rod bearings are available in .010"; .020" and .030".

The connecting rod and piston assembly is removed and installed from the top of the engine.

When the rod is installed in the engine, the offset is away from the nearest main bearing. The oil spray hole in bearing end of rod should be on the follow side or away from camshaft, toward right side of vehicle. Torque wrench reading, 50-55 ft. lbs.

Clearance on crankshaft .0008" to .0023". Total side clearance .005" to .009". See Fig. 21.

Every time a connecting rod is removed from an engine or a new rod is to be installed, it should be checked for alignment on a connecting rod aligning fixture as shown in Fig. 22.

There are different types of connecting rod aligners. Follow the instructions issued by the manufacturer, when checking the connecting rod for twist or bend.

When straightening the rod, twist or bend in the opposite direction more than the original twist or bend then return the rod to true alignment. The rod will then retain correct alignment.

PISTON

The piston is aluminum alloy, "T" slotted, cam ground, tin plated, double ribbed at piston pin

FIG. 21—CONNECTING ROD SIDE PLAY

bosses with a heat insulation groove above top ring. The clearance of the piston to the cylinder bore is .003". Check clearance with .003" feeler gauge ¾ wide; feeler gauge should have from 5 to 10 lbs. pull when being removed, see Fig. 23. The gauge should extend the entire length of the piston on the thrust side which is the opposite side from the "T" slot in the skirt.

Pistons are available in the following over-sizes: .010"; .020" and .030".

If it is ever found necessary to install an oversize piston, the cylinder bore must be honed with a regular cylinder honing tool and the manufacturers instructions should be carefully followed to get a true straight cylinder. Do not try to lap

FIG. 22—CONNECTING ROD ALIGNER

in a new piston using compound, as in so doing it will ruin the tin plating on the piston and cause a scoring or wiping condition of both the piston and cylinder walls. See "Checking Cylinder Bores" and "Cylinder Boring", Pages 28 and 29.

PISTON RINGS

Width of compression rings 3/32". Width of oil control ring 3/16". The upper compression ring is installed with the inside beveled edge up. The face of the lower compression ring is tapered .005". The letters "T-O-P" on the upper edge of the ring indicates how the ring is installed, Fig. 24.

When fitting the rings to the cylinder bores, the end gap is .008"—.013". Fig. 25.

Fitting rings to piston grooves, Fig. 26 and 27. Compression rings .0005"—.001". Oil rings .001"—.0015".

Oversize rings are available in the following sizes: .010"; .020"; .030". Use standard rings up to .010" oversize cylinder bores.

PISTON PIN

The piston pin is anchored in the rod with a lock screw and fitted with a clearance of .0001" to .0009" in piston which is equivalent to a light thumb push fit at room temperature of 70°; pin diameter 13/16" (.8117"). See Fig. 28.

Piston pins are available in oversizes .001"; .002"; .003".

ASSEMBLING CONNECTING ROD TO PISTON

Clamp connecting rod in vise using vise jaw protector shields of a soft metal or two pieces of hardwood on each side of connecting rod three inches from piston pin end.

Start piston pin in piston with groove facing down. Assemble piston to connecting rod with the slot in piston, No. 2, Fig. 29 opposite oil spray hole in bearing end of connecting rod, No. 1. Install piston pin clamp screw.

Center piston on pin and place assembly on connecting rod aligning fixture. Tilt piston to left with piston resting against surface plate. With feeler gauge measure clearance between piston skirt and surface plate. See Fig. 30. Tilt piston to right and check clearance. See Fig. 32. If clearance is within .003" on both left and right positions connecting rod is in alignment. A difference greater than .003" indicates connecting rod is twisted.

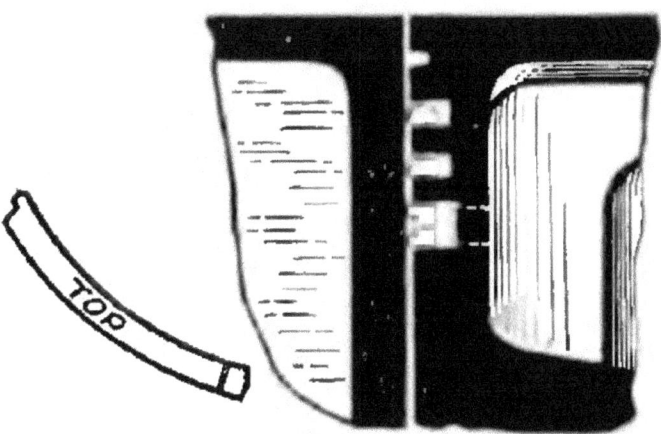

FIG. 24—PISTON RINGS

CHECKING CYLINDER BORES

The best method to be used in determining the condition of the cylinder bores preparatory to reconditioning is the use of a dial gauge such as shown in Fig. 31.

The dial gauge hand will instantly and automatically indicate the slightest variation of the cylinder bores.

To use the dial gauge simply insert in the cylinder bores and move up and down its full length. It is then turned spirally or completely rotated at different points, taking readings at each point. In this manner all variations in the cylinder bores from top to bottom may be determined.

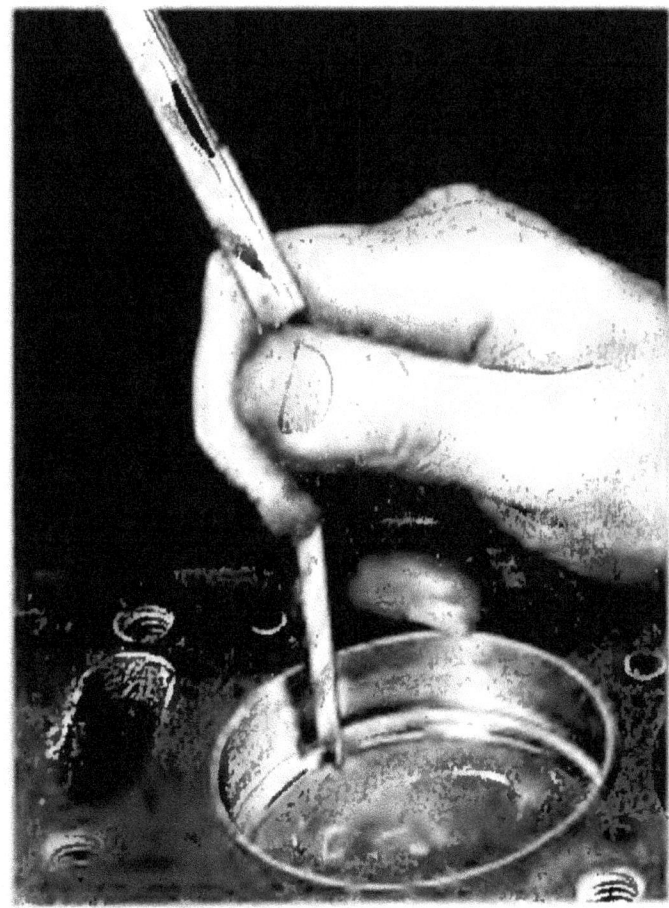

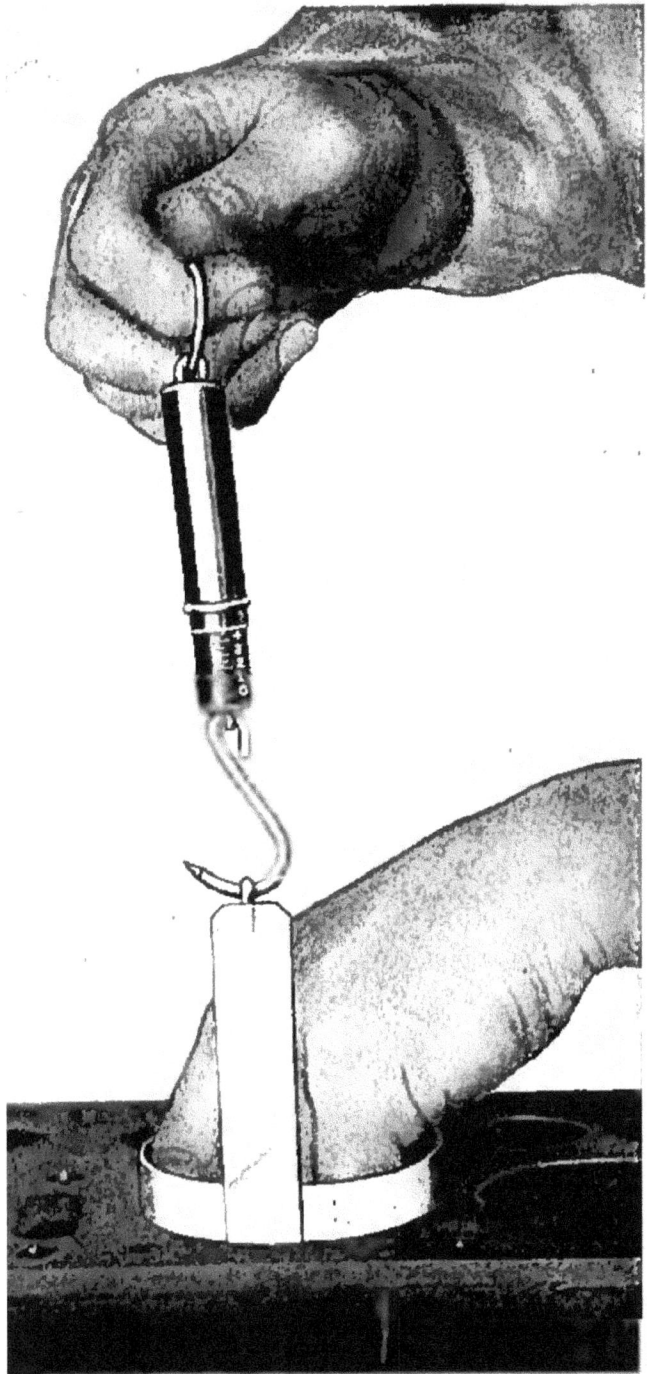

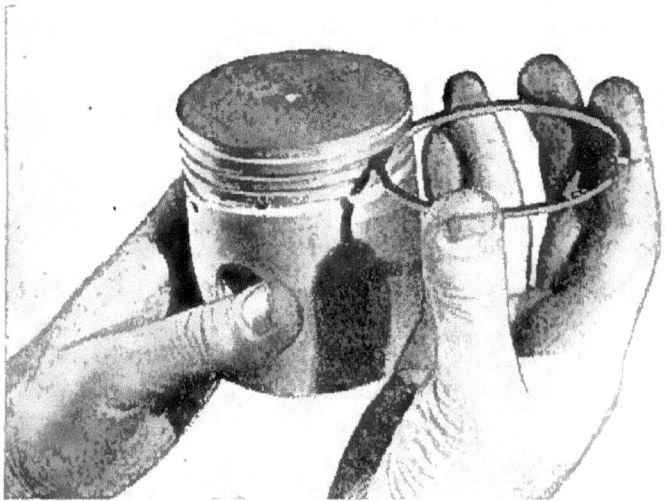

FIG. 26—COMPRESSION RING FITTING

CYLINDER BORING

When cylinders are more than .005" out of true it is best to rebore the cylinders. The instructions furnished by the manufacturer of the equipment should be carefully followed.

After the cylinder has been rebored within .002" of the size desired it should be finished or polished with a cylinder hone. Do not use a piston as a hone. In operating, the hone is placed in the cylinder bore and run up and down the full length of the cylinder wall. This procedure should be followed until the piston can be forced through the bores with a .003" feeler gauge ¾" wide on the thrust side and show a pull on the feeler gauge of five to ten pounds. See Fig. 23.

FIG. 28—PISTON PIN FITTING

OIL PUMP ASSEMBLY

The oil pump is the planetary gear type. It consists of two spur gears enclosed in a one piece housing. It is provided with a relief valve to control maximum oil pressure at all speeds. In operation the oil is drawn from the crankcase through the floating oil intake, Fig. 33. The oil then passes through a drilled passage in crankcase to the oil pump from which it passes to the oil distribution system or drilled passages in crankcase to crankshaft and camshaft bearings.

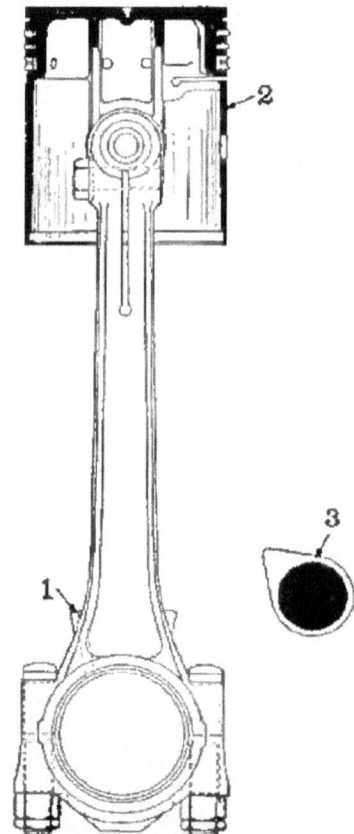

FIG. 27—OIL RING FITTING

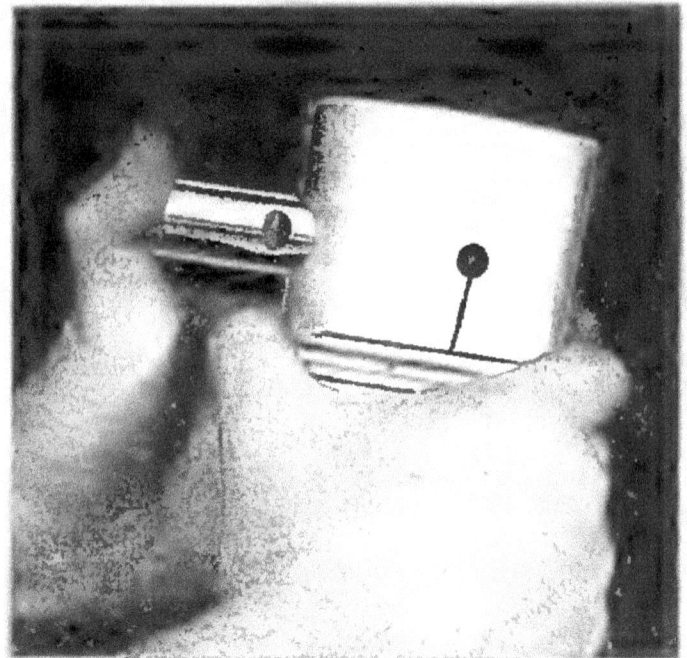

FIG. 29—CONNECTING ROD AND PISTON

The oil pump is driven from a spiral gear on the camshaft which is located at the center of the engine block on left hand side. See No. 22, Fig. 1.

To remove oil pump from engine for dismantling; remove the three nuts on studs holding oil pump to crankcase. Slide oil pump from studs. Remove screw No. 6, Fig. 34, in oil pump cover plate which will allow cover to be removed from housing.

To remove driven gear No. 16, file off one end of straight pin No. 17, with a small drift drive pin through the shaft. The oil pump shaft and rotor No. 12 can be removed from the body in an assembly.

FIG. 31—CYLINDER BORE GAUGE

Set distributor rotor at No. 1 terminal tower in distributor cap and with the points just breaking.

Hold the oil pump in one hand with the oil relief valve retainer in the same position as it

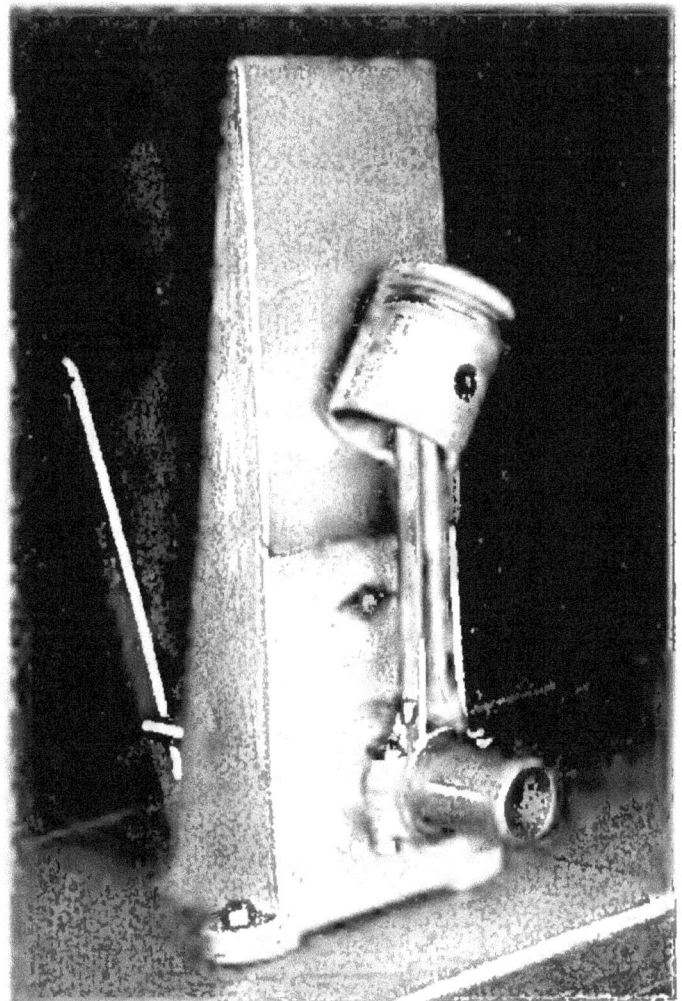

FIG. 30—CONNECTING ROD TWIST-RIGHT

When removing spring retainer No. 1, care must be taken not to lose the small shims No. 3 which govern the spring tension of the relief valve No. 5. Adding shims increases the oil pressure, removing of shims decreases the pressure. The pressure at which the relief valve opens is 40 lbs. actual, however, on instrument panel gauge it will register between 75 and 80. The idle reading should not be less than 10.

When replacing the oil pump on engine the following procedure should be followed in order to have correct timing for the ignition.

Set No. 1 piston coming up on the compression stroke, then turn flywheel so that the timing mark "IGN" appears on the flywheel in the center of the hole in the flywheel housing on the right hand side, Fig. 15.

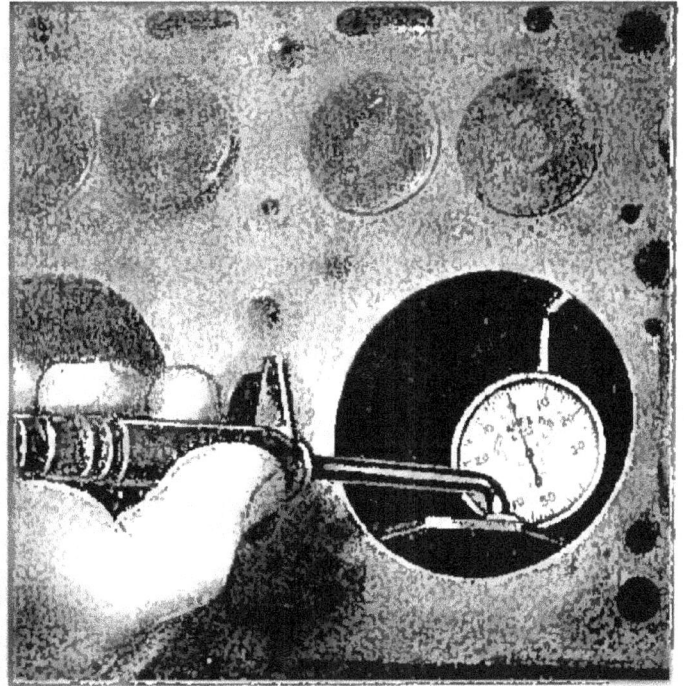

FIG. 32—CONNECTING ROD TWIST-LEFT

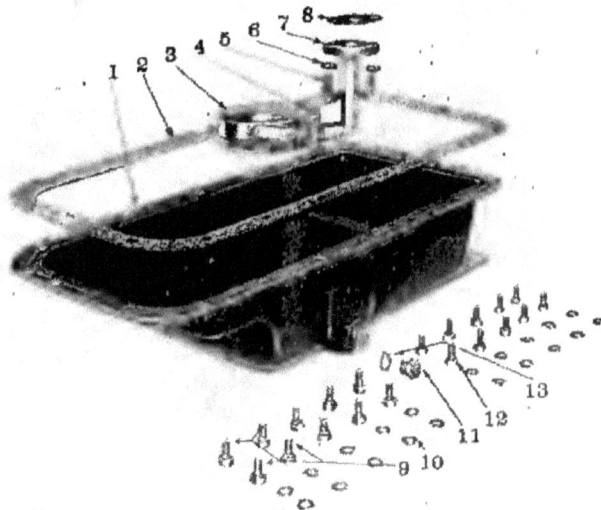

FIG. 33—FLOATING OIL INTAKE AND PAN

No.	Willys Part No.	Ford Part No.	Name
1	A-1167	GPW-6675	Oil Pan Assembly
2	639080	GPW-6710	Oil Pan Gasket
3	630396	GPW-6615	Oil Float Assembly
4	5108	72053-S	Oil Float to Support Cotter Pin
5	636796	355396-S	Oil Float Support to Crankcase Screw
6	51833	34806-S	Oil Float Support to Crankcase—Screw Lockwasher
7	630397	GPW-6617	Oil Float Support
8	630398	GPW-6627	Oil Float Support Gasket
9	51485	20326-S	Oil Pan to Front Engine—Cover Screw
10	51833	34806-S	Oil Pan Screw Lockwasher
11	639979	GPW-6727	Oil Pan Drain Plug
12	51485	20326-S	Oil Pan to Cylinder Block Screw
13	314338	GPW-6734	Oil Pan Drain Plug Gasket

would be when installed in the engine; turn shaft so that the narrow side of slot in driven gear end is toward you, then line up the pin holding driven gear to shaft so that it will fall in line with the right hand side of the slot in pump body. Slide the assembly on studs in the crankcase, feed gear slowly into cam shaft gear, noting when fully set, if the rotor on distributor has moved from its original setting. If so remove oil pump and turn one tooth to obtain the correct setting.

FLOATING OIL INTAKE

The floating oil intake, Fig. 33 is attached to the crankcase with two cap screws. The construction of the float and screen cause it to float on top of the oil, raising and lowering in relation to the amount of oil in the crankcase.

This construction does not permit water or dirt to circulate, which may have accumulated in the bottom of the oil pan, because the oil is drawn horizontally from the top surface.

Whenever removed the float, screen and tube should be cleaned thoroughly in a suitable cleaning fluid to remove any accumulation of dirt.

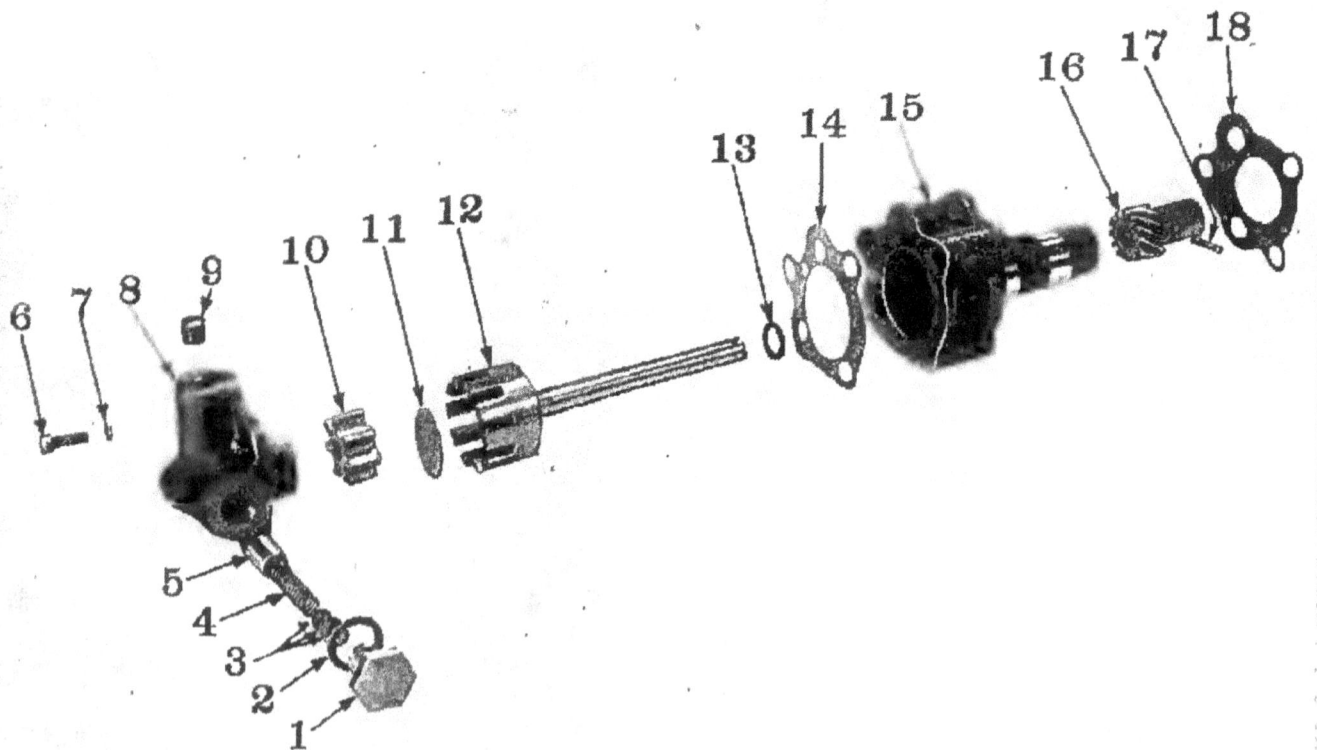

FIG. 34—OIL PUMP

No.	Willys Part No.	Ford Part No.	Name
1	630390	GPW-6644	Oil Pump Oil Relief Spring Retainer
2	634813	GPW-6642	Oil Pump Oil Relief Spring Retainer Gasket
3	630389	GPW-6638	Oil Pump Relief Spring Shim
4	356155	GPW-6651	Oil Pump Relief Plunger Spring
5	630518	GPW-6663	Oil Pump Oil Relief Plunger
6	51819	31070-S	Oil Pump Cover to Body Screw
7	360197		Oil Pump Cover to Body Screw Lockwasher
8	630387	GPW-6664	Oil Pump Cover Assembly
9	52525	353052-S	Oil Pump Cover Plug
10	643306	GPW-6614	Oil Pump Pinion
11	636600	GPW-6673	Oil Pump Rotor Disc
12	636599	GPW-6608	Oil Pump Shaft Assembly
13	375027	GPW-6625	Oil Pump Shaft Gasket
14	639870	GPW-6659	Oil Pump Cover Gasket
15	630384	GPW-6604	Oil Pump Body
16	637435	GPW-6610	Oil Pump Driven Gear
17	630064	GPW-6684	Oil Pump Driven Gear Pin
18	630394	GPW-6630	Oil Pump to Cylinder Block Gasket

FLYWHEEL

The flywheel is made of cast steel, machined all over and balanced to insure smooth engine performance. A steel ring gear is shrunk on the outer edge to mesh with the starter Bendix gear when starting engine.

The flywheel is attached to the crankshaft flange by two dowel bolts and four special head cap screws. When assembling the flywheel to crankshaft, be sure it is properly installed in relation to No. 1 crank throw and that it fits properly to crankshaft flange, to avoid runout or looseness. To check runout use dial indicator attached to the rear engine plate. The runout should not exceed .008" on the rear face near the rim. Torque wrench reading 36-40 ft. lbs.

When installing a new crankshaft or flywheel in service, it is the general practice to replace the tapered dowel bolts with straight snug fitting bolts. The crankshaft and flywheel should be assembled in proper relation, then install the straight bolts previously used and tighten securely. Next use a $\frac{35}{64}$" drill to enlarge the tapered bolt holes and then ream the holes with a $\frac{9}{16}$" (.5625") straight reamer and install two flywheel to crankshaft bolts No. 116295 with nut No. 52804 and lockwasher No. 52330, instead of the two dowel bolts formerly used. This procedure overcomes the difficulty in correctly tapering holes in the field.

OIL FILTER

The oil filter, No. 35 is so designed that it will effectively control contamination of engine oil. The filter element removes particles of dust, carbon and other foreign material from the oil which cause discoloration and sludge.

The inlet line to the filter is connected to the oil distribution line at the front plug on left hand side of engine. The outlet or oil return line to engine connects to the timing chain cover.

When the oil on the level indicator in the engine filler tube becomes dark, remove the oil filter cover; remove the drain plug and drain out the sludge after which replace the drain plug. Next, remove the element and install a new element. Install new cover gasket; reinstall cover, start engine and check for leaks; then check oil level; add to oil supply if necessary.

OIL PRESSURE GAUGE

The oil pressure gauge is of the Bourdon or hydraulic type and measures the pressure of the oil applied to the engine bearings. It does not indicate the amount of oil in the engine crankcase or the need for changing of the engine oil.

A pressure tube connects the gauge unit to the engine. It requires no special attention other than to see that the connection to the engine is tight.

If the unit becomes inoperative, it should be replaced as its construction does not permit repair or adjustment.

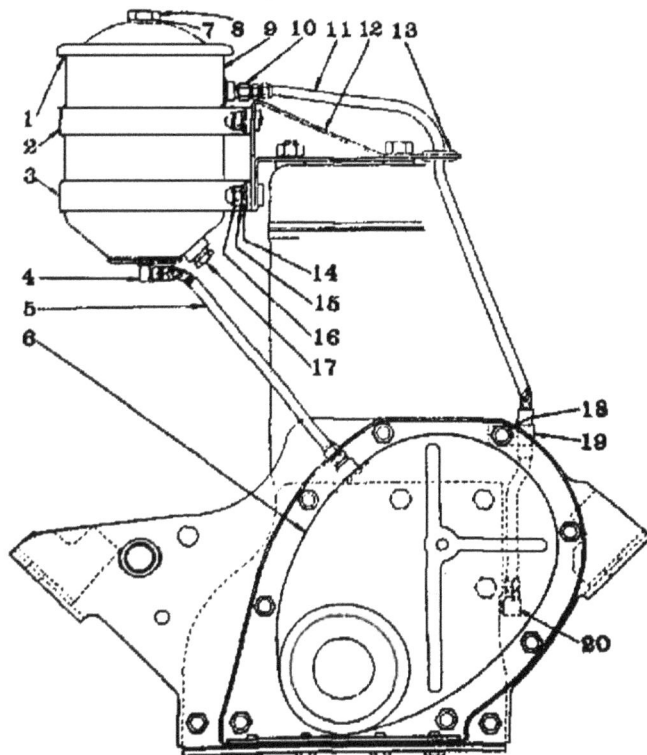

FIG. 35—OIL FILTER

No.	Willys Part No.	Ford Part No.	Name
1	A-1235	GPW-18688	Cover Gasket
2	A-1251	GPW-18658	Clamp Assembly
3	A-1251	GPW-18658	Clamp Assembly
4	384569	GPW-9628	Inverted Flared Tube Ell
5	A-1198	GPW-18666	Oil Filter Outlet Tube
6	A-1190		Chain Cover Assembly
7	A-1233	GPW-18675	Cover Bolt Gasket
8	A-1232	GPW-18691	Cover Bolt
9	A-1230	GPW-18660	Oil Filter Assembly
10	387801	9N-18679	Inverted Flared Tube Connector
11	A-1197	GPW-18667	Oil Filter Inlet Tube
12	A-1247	GPW-18663	Oil Filter Bracket Assembly
13	345961	GPW-13434-A2	Rubber Grommet
14	52274	34746-S2	Plain Washer
15	51833	34806-S	Lockwasher
16	51396	24347-S2	Screw (Filter to Bracket)
17	A-1237	358040-S	Drain Plug
18	5919	24389-S2	Screw (Timing Chain Cover)
19	A-1289	GPW-14585	Inlet Tube Clip
20	384569	GPW-9628	Inverted Flared Tube Ell
—	A-1236	GPW-18632	Filter Element Assembly

ENGINE SUPPORT PLATE AND MOUNTING

The front engine support plate is bolted to the front face of the cylinder block and forms the back panel for the attachment of the timing chain cover.

The rubber engine mountings, which are attached to the frame side rail brackets and to the support plate, prevent fore-and-aft motion of engine yet allow free side wise and vertical oscillation which has the effect of neutralizing vibration at its source, No. 9, Fig. 36.

The rear engine plate is attached to the rear of the cylinder block and provides a means for attaching the flywheel bell housing.

The engine is attached to the center cross member of the frame on a mounting which attaches to the bottom of the transmission. Torque wrench reading, 38-42 ft. lbs.

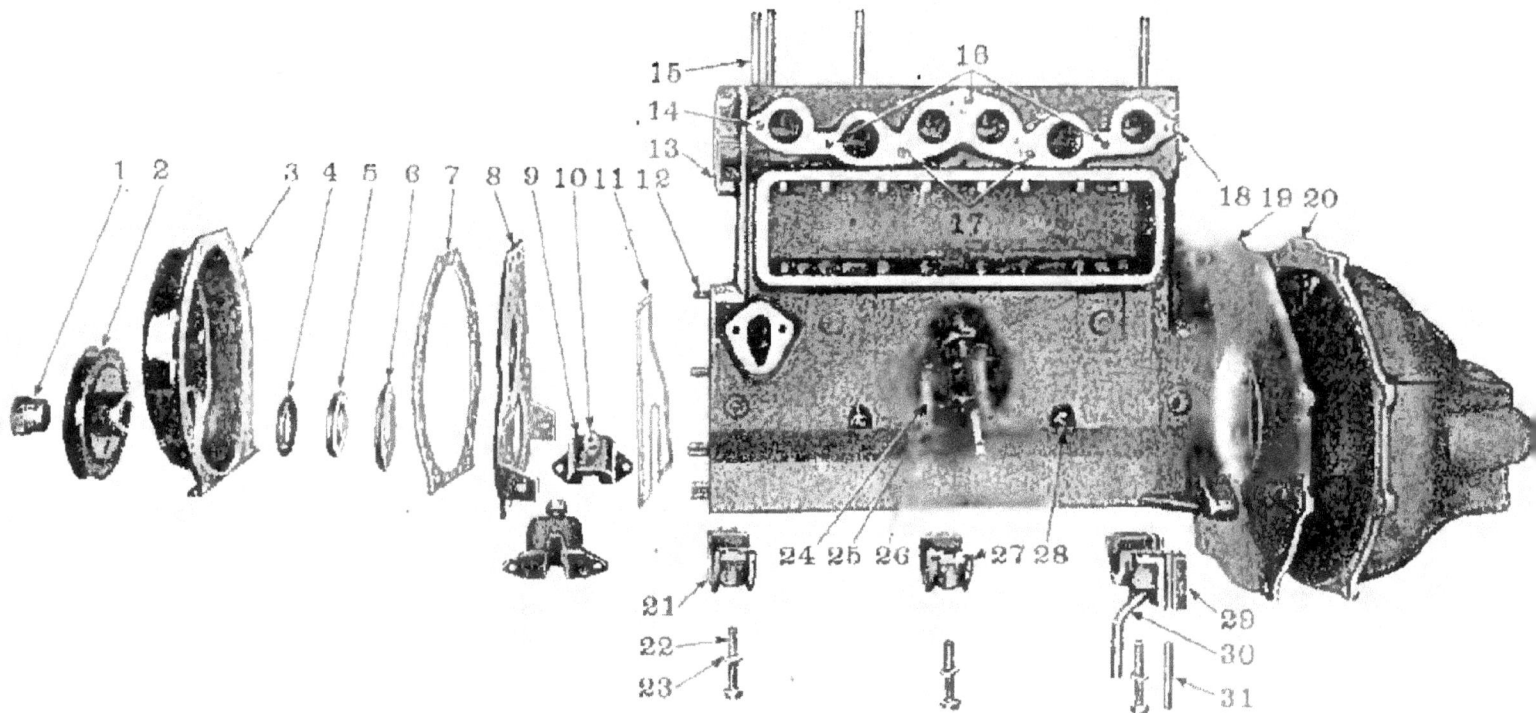

FIG. 36—TIMING CHAIN COVER, CYLINDER BLOCK AND BELL HOUSING

No.	Willys Part No.	Ford Part No.	Name	No.	Willys Part No.	Ford Part No.	Name
1	387633	GPW-6319	Starting Crank Nut Assembly	17	632159	88057-S7	Inlet and Exhaust Manifold Stud—1"
2	638113	GPW-6312	Fan Drive Pulley	18	349712	88082-S	Inlet and Exhaust Manifold Stud—1
3	A-1190		Chain Cover Assembly	19	A-5121	GPW-7007	Engine Plate—Rear
4	637098	GPW-6700	Crankshaft Packing—Front End	20	A-439	GPW-6392	Transmission Bell Housing Assembly
5	375920	GPW-6287	Crankshaft Packing Retainer	21	630285	GPW-6329	Crankshaft Bearing Cap—Front
6	375877	GPW-6310	Crankshaft Oil Slinger	22	381519	GPW-6345	Bearing Caps to Crankcase Screw
7	630365	GPW-6288	Chain Cover Gasket	23	5009	34809-S	Bearing Cap to Crankcase Screw Lock washer
8	A-1463	GPW-6031	Engine Plate Front Assembly	24	375981	88141-S	Oil Pump to Cylinder Block Stud
9	A-542	GPW-6038	Engine Support Insulator—Front	25	51833	34806-S	Oil Pump Stud Nut Lockwasher
10	5916	33846-S	Engine Support Insulator Nut—to Engine Plate	26	5910	33798-S	Oil Pump Stud Nut
11	630359	GPW-6020	Cylinder Block Gasket—Front	27	630288	GPW-6330	Crankcase Bearing Cap—Center
12	384958	88032-S	Chain Cover and Front Plate Stud	28	5085		Oil Passage Plug
13	A-1272		Cylinder Block and Bearings Assembly	29	637236	GPW-6325	Crankshaft Bearing Cap—Rear
14	349712	88082-S	Inlet and Exhaust Manifold Stud—1⅛"	30	630294	GPW-6326	Crankshaft Rear Bearing Drain Pipe
15	349368	GPW-6066	Cylinder Head Stud	31	637790	GPW-6701	Crankshaft Rear Bearing Cap Packing
16	300143	88042-S	Inlet and Exhaust Manifold Stud—1½"				

ENGINE TROUBLES AND CAUSES

Poor Fuel Economy

Ignition Timing Slow or Spark Advance Stuck
Carburetor Float High
Accelerator Pump Not Properly Adjusted
Gasoline Leakage
Leaky Fuel Pump Diaphragm
Loose Engine Mounting causing high gasoline level in Carburetor.
Low Compression
Valves Sticking
Spark Plugs Bad
Weak Coil or Condenser
Improper Valve Tappet Clearance
Carburetor Air Cleaner Dirty
Clogged Muffler or bent Exhaust Pipe

Lack of Power

Low Compression
Ignition System (Timing Late)
Improper Functioning Carburetor or Fuel Pump
Gasoline Lines Clogged
Air Cleaner Restricted
Engine Temperature High
Improper Tappet Clearance
Sticking Valves—Valve Timing
Leaky Gaskets
Muffler Clogged
Bent Exhaust Pipe
Old Gasoline

Low Compression

Leaky Valves
Poor Piston Ring Seal
Sticking Valves
Valve Spring Weak or Broken
Cylinder Scored or Worn
Tappet Clearance Incorrect
Piston Clearance too Large
Leaky Cylinder Head Gasket

Burned Valves and Seats

Sticking Valves or too loose in Guides
Improper Timing
Excessive Carbon around Valve Head and Seat
Overheating
Valve Spring Weak or Broken
Valve Tappet Sticking or not set to .014"
Clogged Exhaust System

Valves Sticking

Warped Valve
Improper Tappet Clearance
Carbonize or scored Valve Stems
Insufficient Clearance Valve Stem to Guide
Weak or Broken Valve Spring
Valve Spring Cocked
Contaminated Oil

Overheating

Defective Cooling
Thermostat Inoperative
Improper Ignition Timing
Improper Valve Timing
Excessive Carbon Accumulation
Fan Belt too Loose
Clogged Muffler or Bent Exhaust Pipe
Oiling System Failure
Scored or Leaky Piston Rings

Popping-Spitting-Detonation

Improper Ignition
Improper Carburetion
Excessive Carbon Deposit in Combustion Chamber
Poor Valve Seating
Sticking Valves—Broken Valve Spring
Tappets Adjusted too Close
Spark Plug Electrodes Burnt
Water or Dirt in Fuel—Clogged Lines
Improper Valve Timing

Excessive Oil Consumption

Piston Rings Stuck in Groove, Worn or Broken
Piston Rings Improperly fitted or Weak
Piston Ring Oil Return Holes Clogged
Excessive Clearance Main and Connecting Rod Bearings
Oil Leaks at Gasket or Oil Seals
Excessive Clearance Valve Stem to Valve Guide (Intake)
Cylinder Bores Scored-Out-of-Round or Tapered
Too Much Clearance Piston to Cylinder Bore
Misaligned Connecting Rods
High Road Speeds or Temperature

Bearing Failure

Crankshaft Bearing Journal Out of Round
Crankshaft Bearing Journals Rough
Lack of Oil—Oil Leakage
Dirty Oil
Low Oil Pressure or Oil Pump Failure
Drilled Passages in Crankcase or Crankshaft Restricted
Oil Screen Dirty
Connecting Rod Bent

ENGINE SPECIFICATIONS

Type..L. Head
  Number of Cylinders..4
  Bore..3⅛"
  Stroke..4⅜"
  Piston Displacement...................134.2 cu. in.
  Compression Ratio.....................6.48 to 1
  Horse Power Max. Brake.................60 @ 4000
  Compression...............111 lbs. @ 185 R.P.M.
  S.A.E. Horse Power........................15.63
  Maximum Torque........105 Lbs.-Ft. @ 2000 R.P.M.
  Firing Order............................1-3-4-2

Cylinder Block:
  Bore Size.............................3.125"—3.127"

Cylinder Head:
  Torque Wrench Pull
  Cylinder Head Screw...............65-75 Lbs.-Ft.
  Cylinder Head Stud Nut............60-65 Lbs.-Ft.

Crankshaft:
  Counter Weights.............................4

Crankshaft Main Bearings:
  Bearing Journals............................3
    Front...............................2.3340"x1.920"
    Center..............................2.3340"x1¹³⁄₁₆"
    Rear................................2.3340"x1¾"

  Thrust......................................Front

  End-play...............................004"-.006"
  Bearing Clearance......................001"
  Type..........................Steel back, babbitt lined
  Non-Adjustable............Replaceable Without Reaming
  Torque Wrench Pull....................65-70 Lbs.-Ft.

Connecting Rod:
  Center to Center Length...................9⅜"
  Upper End................Piston Pin Locked in Rod
  Lower Bearing Type......Steel back, babbitt lined, replaceable
  Lower Bearing Diameter and Length......1¹⁵⁄₁₆"-x1⅝"
  Clearance on Crankshaft..............0008"-.0023"
  Side Clearance.......................005"-.009"
  Torque Wrench Pull...................50-55 Lbs.-Ft.
  Installation................................From Top
    Offset away from nearest main bearing
    Oil spray hole away from camshaft

Piston:
  Lo-Ex Lynite T-Slot—Oval Ground—Tin Plated—Heat dam
  Length...................................3¾"
  Clearance Top Land...................0205"-.0225"
  Clearance Skirt.......................003"
  Oversize pistons Available.....010", .020", .030"
  Number Rings................................3
    Compression Ring...............2-Width ⁵⁄₃₂"
    Oil Ring........................1-Width ³⁄₁₆"
    Ring Gap.......................008"-.013"
    Ring to Groove Clearance.......0005"-.0015"
  Piston Pin Hole
    Diamond Bored..............¹³⁄₁₆" (.8007"-.8119")

Piston Pin:
  Length.................................2²⁵⁄₃₂"
  Diameter...............................¹³⁄₁₆"
  Type...............................Locked in Rod
  Clearance in Piston..............0001"-.0009"
  Oversize Pins Available.......001", .002", .003"

Engine Specifications—Continued

Camshaft:
    Number of Bearings....................................4
    Bearing Journal Diameter:
        Front..................................2 5/16"
        Front Intermediate....................2 1/4"
        Rear Intermediate.....................2 3/16"
        Rear..................................1 3/4"
    Thrust Taken......................................Front
    End Play Control.....................Plunger and Spring

Camshaft Bearings:
    Front....................Steel Back Babbitt Lined
    Clearance.........................002"-.0035"

Intake Valve:
    Tappet Clearance Cold................................014"
    Seat Angle...45°
    Diameter Head.....................................1 17/32"
    Length Over-all....................................5 3/4"
    Stem Diameter.....................................373"
    Stem to Guide Clearance.................0015" to .00325"
    Intake Opens..........9° BTC Flywheel (.039" Piston travel)
    Intake Closes........50° ABC Flywheel (3.772" Piston travel)
    Lift..23/64"

Exhaust Valve:
    Tappet Clearance Cold................................014"
    Seat Angle...45°
    Diameter Head.....................................1 15/32"
    Length Over-all....................................5 3/4"
    Stem Diameter.....................................3725"
    Stem to Guide Clearance................002"-.00375"
    Exhaust Opens.......47° BBC Flywheel (3.799" Piston travel)
    Exhaust Closes........12° ATC Flywheel (.054" Piston travel)
    Lift..23/64"

Valve Spring:
    Free Length.......................................2 1/2"
    Spring Pressure Valve Closed...........50 lbs. length 2 7/64"
    Spring Pressure Valve Open............116 lbs. length 1 3/4"
    Closed Coil End of Spring..........Installed Up against Block

Valve Tappet:
    Overall Length....................................2 7/8"
    Stem Diameter........................6240"-.6245"
    Clearance To Guide...................0005"-.002"
    Adjusting Screw................3/8"-24 thd. x 1 1/32"

Timing Chain:
    Link-Belt
    Number Links......................................47
    Width..1"
    Pitch..1/2"
    Type.......................................Non-adjustable

Fan Belt:
    Type.."V"
    Angle of Vee......................................42°
    Length outside...................................44 1/8"
    Width...1 1/16"

Oil Pump:
    Type.......................................Planetary Gear
    Driven from Camshaft..............................Gear

Oil Pressure Relief:
    Pressure 40 lbs. actual—75 gauge at 30 miles per hour.
    Adjustable..................Shims in Spring Retainer.

Oil Filter............................Purolator No. 27078

Spark Plugs............................Champion QM-2

CLUTCH

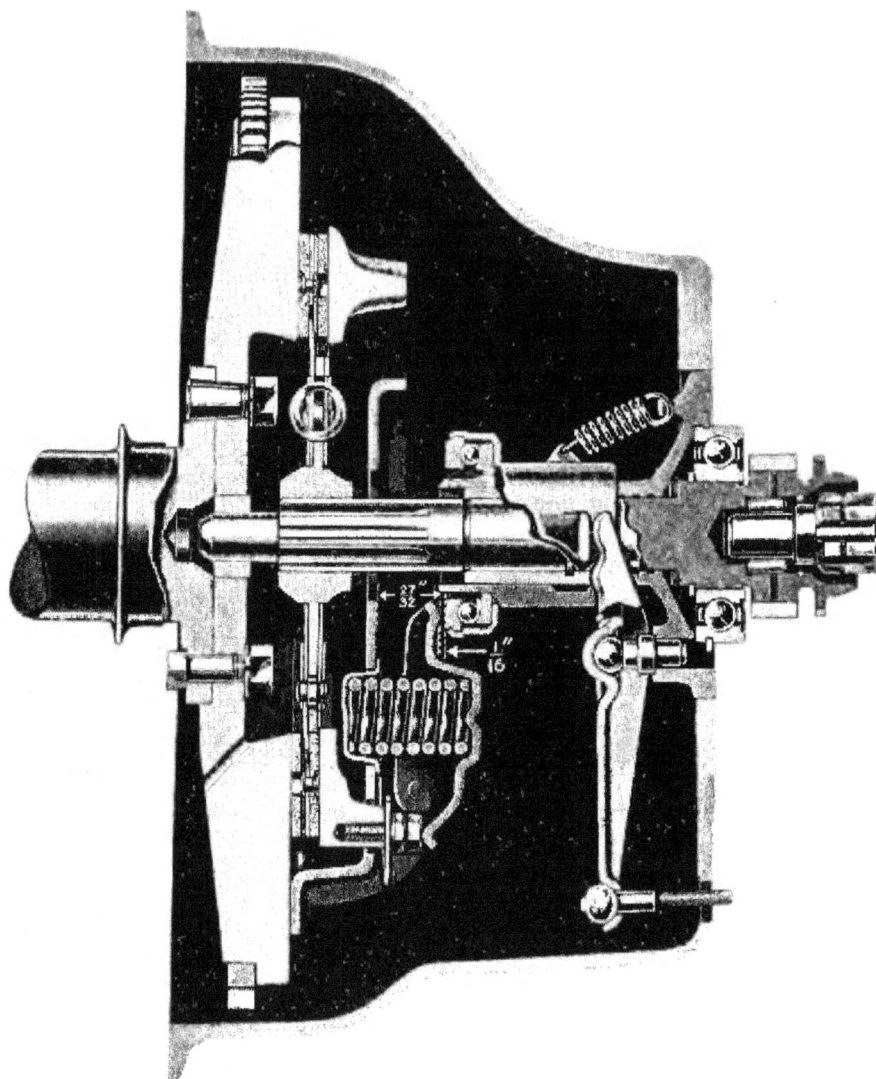

FIG. 1—CLUTCH SECTIONAL VIEW

The clutch is a single plate 8″ dry disc type. The driven plate has a spring center vibration neutralizer, two facings ⅛″ thick, outside diameter 7⅞″, inside diameter 5⅛″.

There are two adjustments, the free pedal play which should be maintained at ¾″ and the other, the pressure plate finger adjustment. See Fig. 1.

As the clutch facings wear it diminishes the free pedal travel. When clutch pedal rests tightly against the toe board it is necessary to adjust the clutch cable.

Clutch Pedal Adjustment

Lengthening or shortening clutch control lever cable No. 11, Fig. 2 governs the clearance of the clutch release bearing to clutch fingers, Fig. 1, which should be maintained at 1/16″. This represents ¾″ free pedal travel. This also disengages the clutch release bearing and prevents unnecessary wear while the engine is running.

Loosen Clutch control lever cable adjusting yoke locknut, No. 24, Fig. 2. With a wrench unscrew cable No. 11, then tighten locknut.

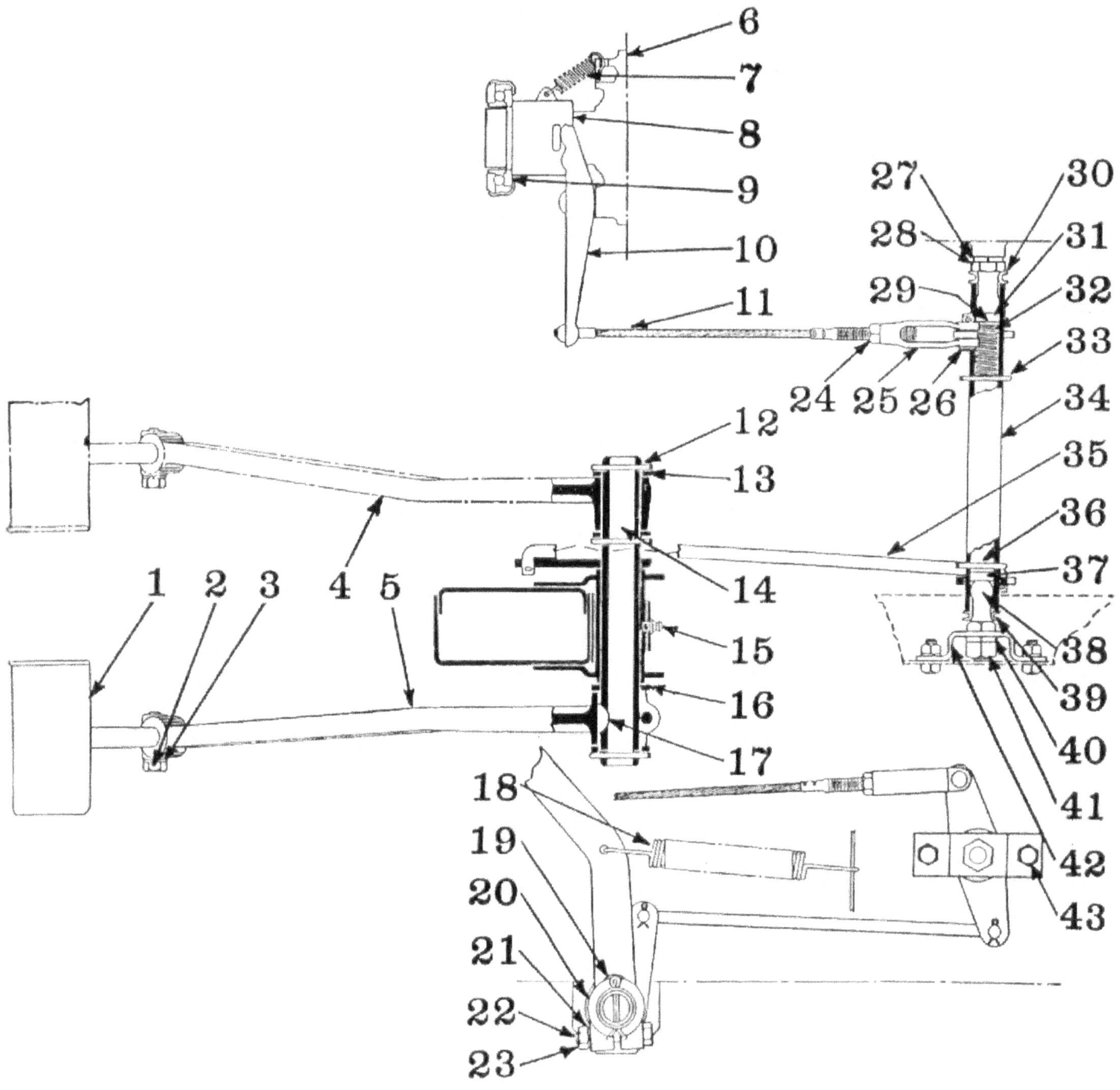

FIG. 2—CLUTCH CONTROL

No.	Willys Part No.	Ford Part No.	Name	No.	Willys Part No.	Ford Part No.	Name
1	A-1360	GPW-7525	Clutch Pedal Pad Assembly	23	5910	33798-S	Clutch Pedal Clamp Bolt Nut
2	6157	24426-S	Clutch Pedal Clamp Bolt	24	5910	33798-S	Clutch Control Lever Cable Yoke End Locknut
3	51833	34806-S	Clutch Pedal Clamp Bolt Lockwasher	25	632177	GPW-7532	Clutch Control Lever Cable Yoke End
4	A-1386	GPW-2452	Brake Pedal	26	339043	73880-S	Clutch Control Lever Cable Clevis Pin
5	A-405	GPW-7520	Clutch Pedal	27	5059	34808-S	Clutch Control Ball Stud Lockwasher
6	640017	GPW-7050	Transmission Main Drive Gear Bearing Retainer	28	A-181	GPW-7514	Clutch Control Ball Stud
7	630117	GPW-7562	Clutch Release Bearing Carrier Spring	29	A-177	GPW-7517	Clutch Control Tube Washer
8	639654	GPW-7561	Clutch Release Bearing Carrier	30	A-887	GPW-7512	Clutch Control Tube Dust Washer
9	635529	GPW-7580	Clutch Release Bearing	31	A-176	GPW-7539	Clutch Control Tube Felt Washer
10	630112	GPW-7515	Clutch Control Lever	32	A-178	GPW-7545	Clutch Control Spring
11	A-5102	GPW-7530	Clutch Control Lever Cable	33	5108	72053-S	Clutch Control Tube Spring Cotter Pin
12	52944	72063-S	Pedal Shaft Cotter Pin	34	A-1355	GPW-7503	Clutch Control Lever & Tube Assembly
13	A-1354	GPW-2138	Master Cylinder Tie Bar	35	A-499	GPW-7521	Clutch Control Pedal Rod
14	A-495	GPW-2473	Pedal Shaft Assembly	36	5108	72053-S	Clutch Control Tube Cotter Pin
15	638792	359043-S7	Pedal Shaft Hydraulic Grease Fitting	37	A-176	GPW-7539	Clutch Control Tube Felt Washer
16	A-498	356561-S	Pedal Shaft Washer	38	A-181	GPW-7514	Clutch Control Ball Stud
17	5036	74178-S	Clutch Pedal to Shaft Woodruff Key	39	A-887	GPW-7512	Clutch Control Tube Dust Washer
18	630593	GPW-7523	Clutch Pedal Retracting Spring	40	5059	34808-S	Clutch Control Ball Stud Lockwasher
19	52944	72063-S	Pedal Shaft Cotter Pin	41	5336	33801-S	Clutch Control Ball Stud Nut
20	A-498	356561-S	Pedal Shaft Washer	42	A-130	GPW-7509	Clutch Control Frame Bracket
21	51833	34806-S	Clutch Pedal Clamp Bolt Lockwasher	43	52132	24327-S	Clutch Control Frame Bracket Screw
22	50992	24505-S	Clutch Pedal Clamp Bolt				

Reconditioning

When it is necessary to recondition the clutch, follow the procedure outlined under "Transmission" for the removal of transmission and transfer case from Engine. Remove the bell housing, then the following procedure is suggested in disassembling the clutch.

Prick punch both the pressure plate and the flywheel so that the assembly can be installed in the same position after the repairs are performed.

The cap screws holding the clutch pressure plate to the flywheel should be loosened in sequence a little at a time so as to prevent distortion of the clutch bracket No. 10, Fig. 3. It is advisable where the clutch facings have worn that a new driven plate assembly be used in preference to relining the old plate as in many cases it is found that the torque springs in the hub of the driven plate have somewhat weakened or the center plate No. 13

To assemble the clutch to the flywheel first put a small amount of light cup grease in the clutch shaft bushing; install driven plate, with short end of hub towards the flywheel then place pressure plate assembly in position. With clutch pilot arbor or a clutch shaft, align driven plate leaving arbor in place while tightening the pressure plate cap screws. Remove clutch pilot arbor and then adjust clutch fingers.

Adjustment of Clutch Fingers

Adjustment of the clutch fingers is accomplished by loosening the lock nut on adjusting screws, then turning screws clockwise or counter-clockwise until the measurement from the face of the fingers (release bearing contacts) to the face of the clutch bracket measures $^{27}\!/_{32}''$. Fig. 5. All fingers must be adjusted so that clutch release bearing will contact the three fingers simultaneously.

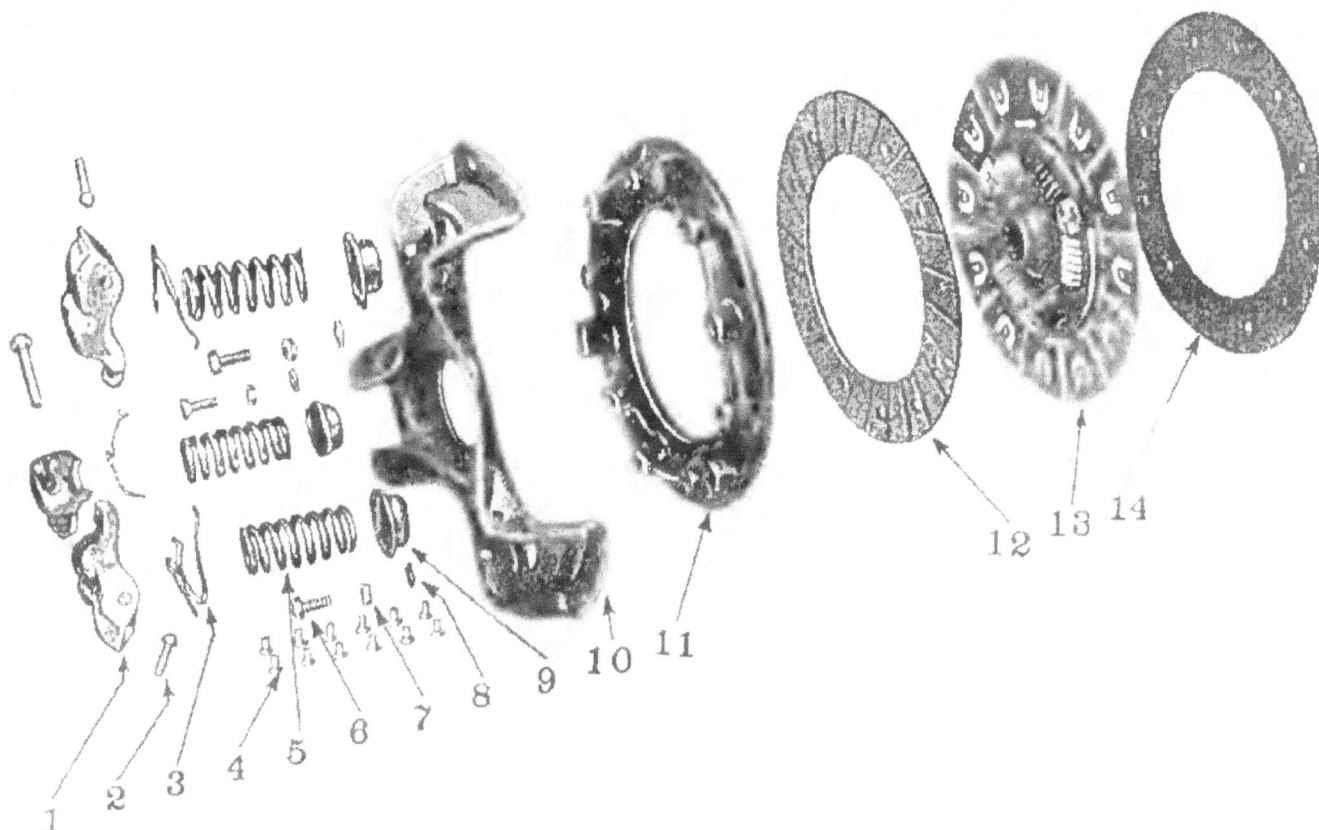

FIG. 3—CLUTCH

has become distorted and if relined would not give satisfactory performance.

The clutch pressure plate fingers are properly adjusted before vehicle leaves the factory and should not require adjustment excepting where it is necessary to install new springs, fingers or pressure plate. When new parts are installed, it will be necessary to readjust the fingers, which adjustment can be made after the assembly is attached to the flywheel.

No.	Willys Part No.	Ford Part No.	Name
1	638158	CPW-7591	Clutch Lever
2	638159	GPW-7585	Clutch Lever Pivot Pin
3	638153	GPW-7590	Clutch Pressure Plate Return Spring
4	374681	351926-S	Clutch Facing Rivet—Tubular
5	638993	GPW-7572	Clutch Pressure Spring
6	638154	24325-S	Clutch Adjusting Screw
7	638155	33921-S	Clutch Adjusting Screw Lock Nut
8	638305	34745-S	Clutch Adjusting Screw Washer
9	638157	GPW-7567	Clutch Pressure Spring Cup
10	638151	GPW-7570	Clutch Bracket
11	638152	GPW-7566	Clutch Pressure Plate
12	636778		Clutch Facing—Rear
13	638755	GPW-7550	Clutch Driven Plate and Hub (with Facings)
14	371567	GPW-7549	Clutch Facing—Front

Next assemble the bell housing to the engine.

The clutch release bearing No. 9, Fig. 2 is pre-lubricated and the lubricant lasts the life of the bearing. If the bearing is rough, a new bearing should be installed.

Make sure that the clutch release bearing carrier return spring No. 7, Fig. 2 is hooked into place. For the balance of the assembly, reverse the operations that were used in the disassembly, referring to instructions given under "Transmission." Finally adjust the clutch release cable so there is ¾" free pedal travel before release bearing contacts the fingers.

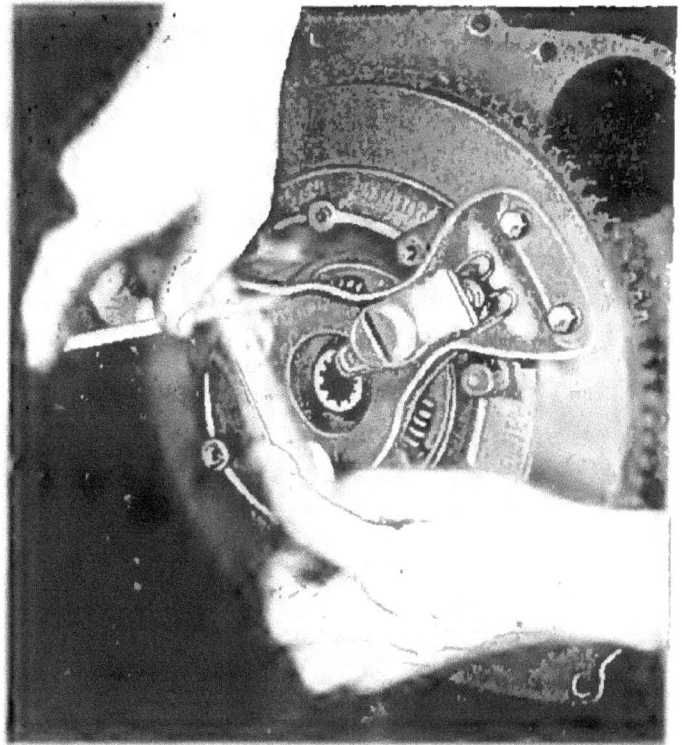

FIG. 5—CLUTCH FINGER ADJUSTMENT

FIG. 4—CLUTCH PRESSURE PLATE REMOVAL

Clutch Pressure Plate

When it is found necessary to install a new clutch pressure plate or clutch spring dismantle the assembly as follows: See Fig. 4.

Select a board or a surface plate of sufficient length and width to support the pressure plate bracket at all points. Place board or surface plate on arbor press table. Place a piece of wood 2½" square on top of clutch fingers, then depress springs, holding down while the adjusting screws are removed from pressure plate.

After screws No. 6, Fig. 3 are removed, the clutch pressure plate return springs, No. 3, may be removed.

Release press arbor slowly to prevent clutch pressure springs, No. 5, flying out from under the clutch fingers No. 1.

In checking assembly make sure that the clutch pressure spring cups, No. 9 are assembled in bracket No. 10 with indentation in bottom towards the center.

The assembly of the pressure plate is the reverse of operations used in dismantling.

CLUTCH TROUBLES AND REMEDIES

SYMPTOMS	PROBABLE REMEDY
Slipping:	
Improper Adjustment	Adjust Pedal Free Travel ¾"
Weak Pressure Springs	Replace
Lining Oil Soaked	Install New Driven Plate
Worn Linings or Torn Loose from Plate	Install New Driven Plate
Burned Clutch	Replace
Grabbing:	
Gummy or Worn Linings	Install New Driven Plate
Loose Engine Mountings	Tighten
Scored Pressure Plate	Install New Plate
Improper Clutch Lever (Finger) Adjustment	Readjust
Clutch Plate Crimp or Cushion Flattened Out	Replace Driven Plate
Dragging:	
Too much Pedal Play	Adjust
Improper Finger Adjustment	Readjust
Pressure Plate Binds in Bracket	Adjust
Warped Pressure or Driven Plate	Replace
Torn Clutch Facing	Replace
Rattling:	
Broken or Weak Return Springs in Driven Plate	Replace
Worn Throw Out Bearing	Replace
Fingers Improperly Adjusted	Readjust
Worn Driven Plate Hub or Clutch Spline Shaft	Replace
Pilot Bushing in Flywheel Worn	Replace

CLUTCH SPECIFICATIONS

Type.................Single, Dry Plate

Driven Plate:

    Make.......................Borg & Beck
    Size........................7⅞"
    Facings......1 Woven and 1 Molded Asbestos
        Diameter.........Inside 5⅛"-Outside 7⅞"
        Thickness.....................⅛"-(.125")
        Torque Capacity.............132 Lbs.-Ft.

Pressure Plate:

    Make...............................Atwood
    Number Springs...........................3
    Spring Pressure at 1⁹⁄₁₆"..........220-230 lbs.

Clutch Release Bearing:

    Type......Sealed Ball Bearing—Prelubricated

Clutch Shaft Bushing:

    Location.....................In Crankshaft
    Material.Bronze Bushing (Impregnated with
        graphite)
    Size........................(Inside) .628"

Clutch Pedal Adjustment:

    Pedal Adjustment..¾" Free Pedal Travel Be-
        fore Release Bearing Con-
        tacts Clutch Fingers.

FUEL SYSTEM

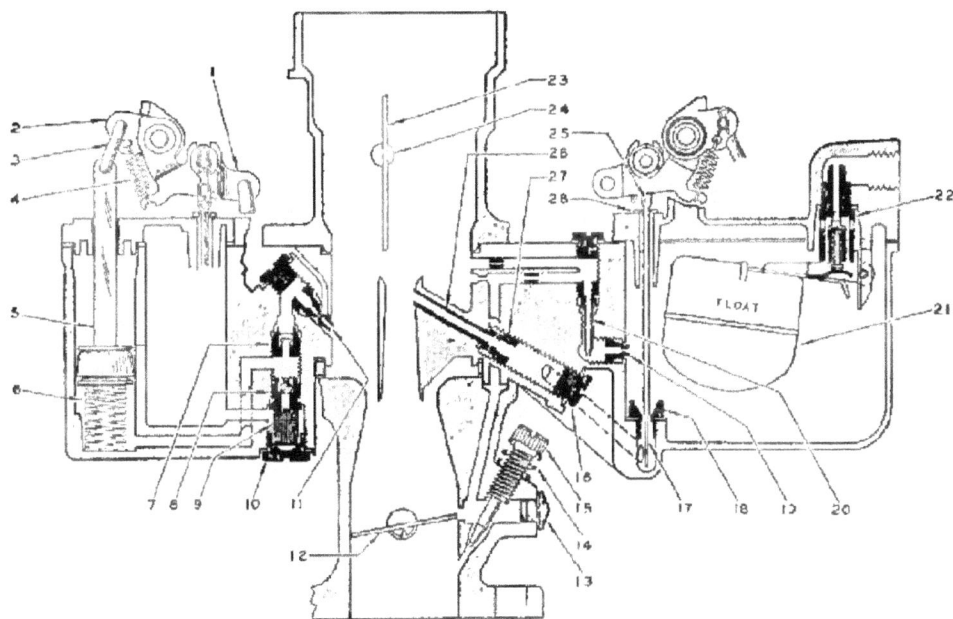

FIG. 1—CARBURETOR

No.	Willys Part No.	Ford Part No.	Name
1	116537	GPW-9529	Pump Operating Lever Assembly
2	116181	GPW-9528	Pump Arm and Collar Assembly
3	116199	GPW-9527	Pump Connecting Link
4	116187	GPW-9570	Pump Arm Spring
5	116195	GPW-9631	Pump Plunger and Rod Assembly
6	116188	GPW-9636	Pump Plunger Spring
7	116204	GPW-9594	Discharge Disc Check Assembly
8	116205	GPW-9576	Intake Ball Check Assembly
9	116175	GPW-9575	Pump Check Strainer
10	116153	GPW-9696	Pump Check Strainer Nut
11	116180	GPW-9940	Pump Jet
12	116154	GPW-9585	Throttle Valve
13	116162	GPW-9579	Idle Port Rivet Plug
14	116183	GPW-9578	Idle Adjustment Screw Spring
15	116176	GPW-9541	Idle Adjustment Screw
16	116164	GPW-9928	Pump Jet also Nozzle Passage Plug and Gasket Assembly
17	116540	GPW-9906	Metering Rod
18	116541	GPW-9914	Metering Rod Jet and Gasket Assembly
19	116179	GPW-9544	Idle Well Jet
20	116539	GPW-9533	Low Speed Jet Assembly
21	116172	GPW-9550	Float and Lever Assembly
22	116174	GPW-9567	Needle, Pin, Spring and Seat Assembly
23	116157	GPW-9549	Choke Valve Assembly
24	116545	GPW-9546	Choke Shaft and Lever Assembly
25	116538	GPW-9907	Metering Rod Spring
26	116166	GPW-9922	Nozzle
27	116161	GPW-9562	Nozzle Retainer Plug
28	116206	GPW-9905	Metering Rod Disc

The Fuel System, Fig. 2, consists of the Fuel Tank, Fuel Lines, Fuel Filter, Fuel Pump, Carburetor and Air Cleaner.

The most important attention necessary to the fuel system is to keep it clean and free of water.

It should be periodically inspected for leaks.

The fuel tank capacity is given on Page 3. The tank sets in a sump in the floor pan and two drain holes are incorporated in this sump to allow for flushing. When the vehicle leaves the factory a cap is placed over the front drain hole to keep out stones and dirt and another is placed in the glove compartment. Should maneuvers in water be necessary, install the second cap over the rear drain hole from the left side of the vehicle. After passing through the water remove cap and return it to the glove compartment.

CAUTION—Whenever the vehicle is to be stored for an extended period, the fuel system should be completely drained. The engine started and allowed to run until carburetor is emptied. This will avoid oxidation of the gasoline, resulting in the formation of gum in the units of the Fuel System.

Information pertaining to the operation and servicing of the units contained in fuel system are covered in the succeeding paragraphs.

Carburetor

The Carter Carburetor, Model WO-539S, Fig. 1, is the plain tube type with a throttle operated accelerator pump and economizing device.

Since carburetion is dependent in several ways upon both compression and ignition, it should always be checked last in an engine tune-up.

The carburetor delivers the proper fuel and air ratios for all speeds of the engine. By proper cleaning and replacing all worn parts, the carburetor will function correctly.

The carburetor can be divided into five circuits which are:

1. Float Circuit
2. Low Speed Circuit
3. High Speed Circuit
4. Pump Circuit
5. Choke Circuit

By treating each circuit separately, the study and repair of the carburetor is made much easier.

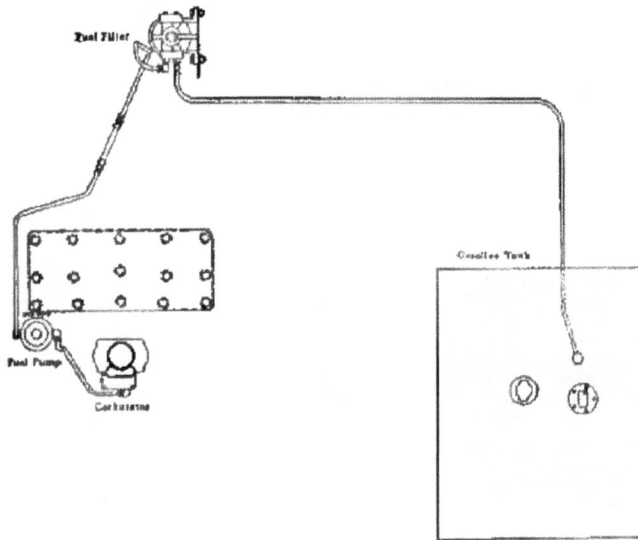

FIG. 2—FUEL SYSTEM

Float Circuit or Fuel Level

The float circuit Fig. 3, is important because it controls the height of the fuel level in the bowl and in the nozzle. If the fuel level is too high, it will cause trouble in the low and the high speed circuits.

The float bowl No. 3, acts as a reservoir to hold a supply of fuel. The level of the fuel in the bowl is controlled by a combination of parts which are: float and lever assembly No. 2, float bowl cover No. 4, needle valve and seat assembly No. 1.

Low Speed Circuit

The idle or low speed circuit, Fig. 4, controls the supply of fuel to the engine during idle and light load speeds up to approximately 20 miles per hour, and it feeds a small amount of fuel during the

entire operation of the high speed circuit (gradually decreasing as speed is increased, above 20 m.p.h.).

During idling and low speed operation of the engine, fuel flows from the float bowl through the idle jet No. 8, to the point where it combines with a stream of air coming in through the by-pass, No. 9. The combining of the air with the fuel atomizes or breaks up the fuel into a vapor.

This mixture of air and fuel continues on through the economizer No. 10 until it begins to pass the point where it is further combined with a stream of air coming in through the lower bleed No. 11. This mixture of fuel and air then flows downward to the idle port chamber and thence into the engine at the port No. 12 and through the idle adjusting screw seat just below. This mixture is richer than the engine requires but when mixed with the air coming past the throttle valve it forms a combustible mixture of the right proportion for idle speeds.

The idle port is slotted so that as the throttle valve is opened it will not only allow more air to come in past it, but will also uncover more of the idle port allowing a greater quantity of fuel and air mixture to enter the intake manifold.

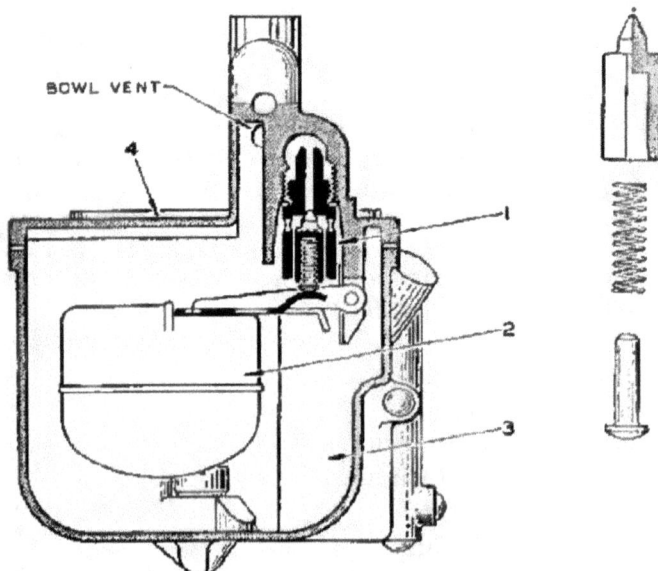

FIG. 3—FLOAT CIRCUIT

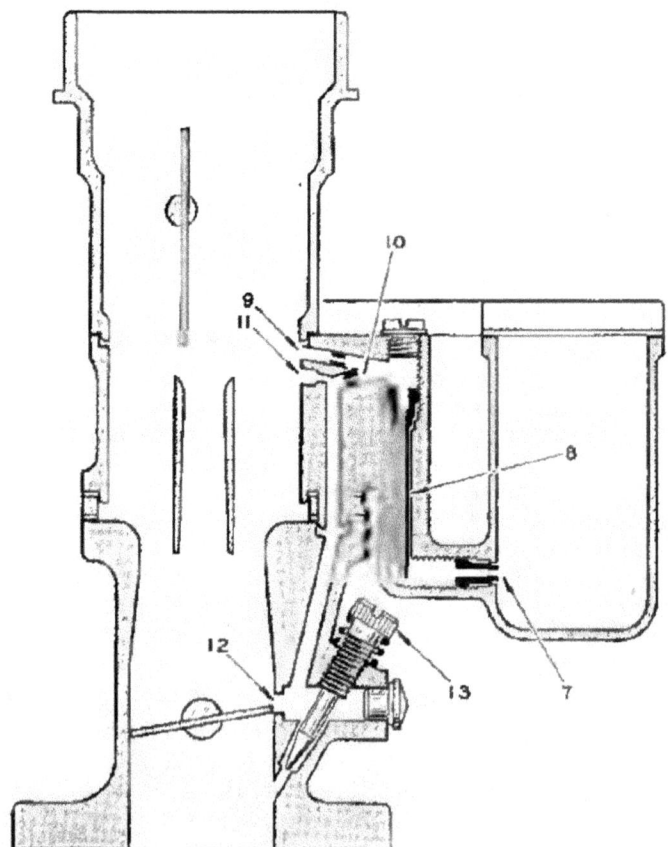

FIG. 4—LOW SPEED CIRCUIT

When the idle speed position of the throttle is fixed at an idle speed of 8 miles per hour, it leaves enough of the slotted port as reserve to cover the range in speed between idle and the time when the high speed system begins to cut in.

The idle adjusting screw No. 13 varies the quantity of the idle mixture.

High Speed Circuit

The high speed circuit, Fig. 5 cuts in as the throttle is opened wide enough for a speed of a little more than 20 miles per hour. The velocity of the air flowing down through the carburetor throat creates a pressure slightly less than atmospheric pressure at the tip of the main nozzle, No. 20.

Since the fuel in the float bowl is acted upon by atmospheric pressure, the difference in pressure between the two points causes fuel to flow from the bowl through the metering jet and out the main nozzle into the throat of the carburetor.

At higher speeds the area of the opening between the jet No. 17 and the metering rod No. 16 governs the amount of fuel going into the engine. At top speeds, the smallest section of the rod is in the jet.

Accelerating Pump Circuit

As the accelerator pedal is depressed, the pump plunger and lever are forced downward. This causes the fuel to leave the cylinder; closes the intake check valve No. 29, Fig. 6, opens the discharge check valve No. 30, and forces fuel into the throat of the carburetor at No. 33.

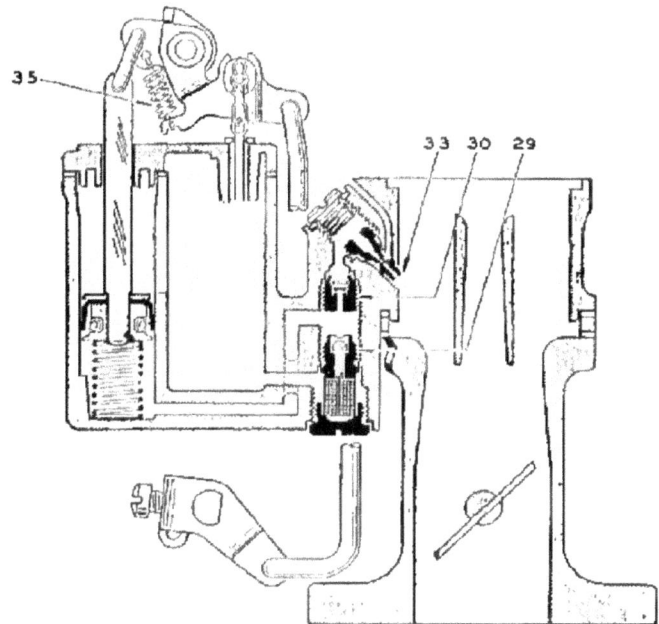

FIG. 6—PUMP CIRCUIT

The action is prolonged by the pump arm spring, No. 35, Fig. 6, since the hole in the top of the pump jet No. 33 restricts the flow of fuel so long as it is being forced out by the pump cylinder. The prolonging of the pump discharge gives the fuel in the high speed circuit sufficient time to flow fast enough to satisfy the demands of the engine.

As the accelerator pedal is allowed to return to its original position, the pump plunger is lifted upward. This creates a reduced pressure in the pump cylinder which opens the intake check valve No. 29 and closes the discharge check valve No. 30, thereby drawing in a new charge of fuel from the bowl.

Choke Circuit

This circuit, Fig. 7 is used only in starting and the warming up of the engine, by reducing the amount of air allowed to enter the carburetor and, thereby producing a richer mixture. It consists of the choke shaft and lever assembly No. 39, choke operating lever and spring No. 40, choke valve No. 37, and screws No. 38.

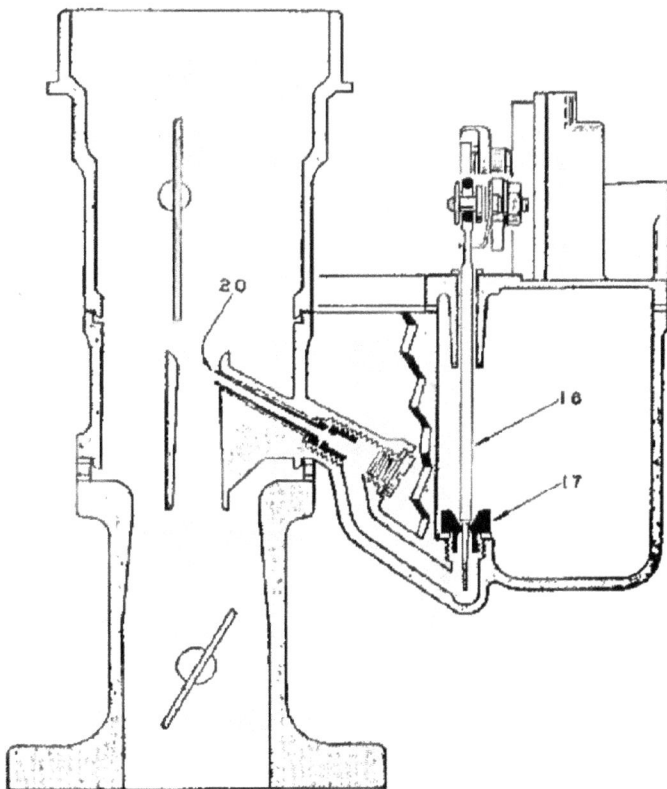

FIG. 5—HIGH SPEED CIRCUIT

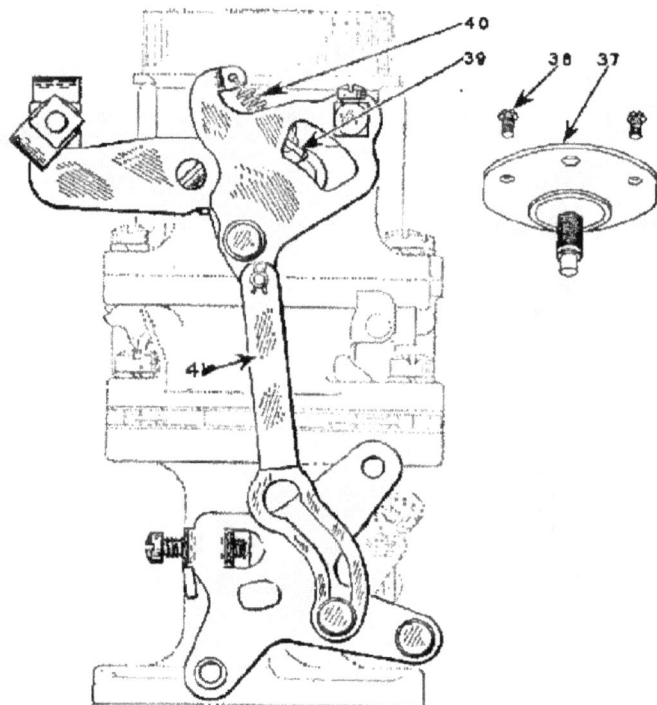

FIG. 7—CHOKE CIRCUIT

SERVICING AND ADJUSTMENT

Float Circuit

The Float Circuit is illustrated in Fig. 3.

If float is loaded with fuel or damaged, or if the holes for the pin are worn, the carburetor will flood. Poor action of float needle results if the lip of the float bracket is worn. In this event, it should be smoothed with emery cloth.

The needle and seat may leak because of wear, damage or sticking and will cause the carburetor to flood. Needles and seats are available only in matched sets. Never replace the needle without replacing the seat.

In determining the float level, Fig. 8, first turn the bowl cover gasket around and with the bowl cover in the position as shown, the float by its own weight, should rest at (⅜") as indicated by the gauge.

To make a change in the float level, it is best to press down with a screw driver on the brass lip of the float, holding up on the float while assembled to the cover of the carburetor. Bending the lip in this way allows it to retain its curvature which is necessary for the correct operation of the float valve.

Be sure the spring and pin in valve are in position and that the spring has not been stretched.

Low Speed Circuit

In the low speed circuit Fig. No. 4 it will be found that the fuel for the low speed circuit does not come through the main metering jet, but through the well jet No. 7, and the low speed jet No. 8, the openings of which are carefully calibrated, so if they are damaged or worn they should be replaced. The jets should always be tightly seated.

The by-pass and air bleed holes, No. 9 and 11, may be restricted. Carbon deposit which forms in the throat of the carburetor may restrict the air bleed holes to the extent that insufficient air will be supplied to mix the fuel before it reaches the idle port, No. 12.

This condition will generally be indicated if it is necessary to screw the idle mixture adjusting screw, No. 13, in closer than the minimum limit of ½ turn. If the condition is bad, a rolling idle may continue even after the idle mixture adjusting screw is screwed entirely in against the seat. These air bleed holes may be cleaned with a soft copper wire.

The idle port must be clean and unrestricted. If it is damaged, the engine will not perform properly at low speeds and a new casting will be necessary.

A letter "C" enclosed within a circle is stamped on the face of the throttle valve. When installed in the carburetor, this side should be toward the idle port, and facing the intake manifold as viewed from the bottom.

To properly center the valve in the throat of the carburetor, the screws should be started in the shaft, and then with the valve tightly closed, (throttle lever adjusting screw backed out), it should be lightly tapped. This will centralize the valve in the bore. Pressure should then be maintained with the fingers until the screws are tightened.

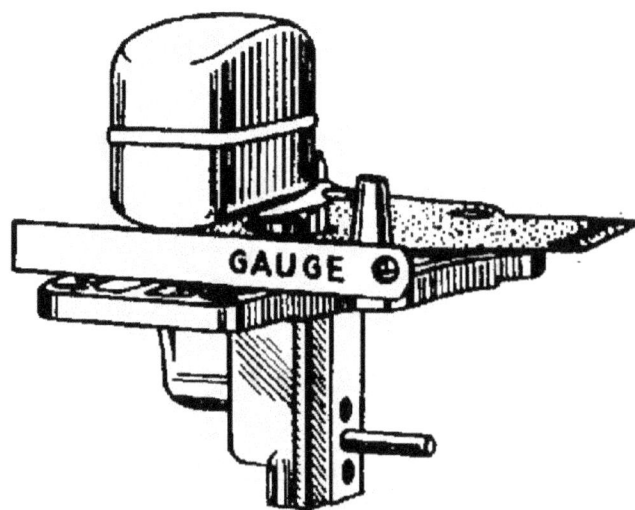

FIG. 8—FLOAT LEVEL SETTING

If the carburetor bore is restricted with carbon deposit it will be necessary to open the throttle wider than the specified opening to obtain the proper idle speed. Opening the throttle more than the specified amount in order to obtain the proper idle will then uncover more of the slotted idle port than was intended. This will result in leaving an insufficient amount of the idle port as a reserve to cover the period between idle and 20 miles per hour, where the high speed system begins to cut in. A flat spot on acceleration will result. Clean by scraping or with emery cloth.

High Speed Circuit

It is rarely necessary to remove the main nozzle No. 20, Fig. 5. It can usually be cleaned by removing plug and blowing out with compressed air. If it is damaged and requires replacing make sure, upon installation, only one gasket is between nozzle and its seat in the casting.

If the carburetor has been in service for a long time or has been tampered with, it may be found the metering rod is improperly adjusted or worn. A worn metering rod will have the effect of a rich mixture above 20 miles per hour. If the metering rod is worn, the metering rod jet will also be worn and both should be replaced.

To adjust metering rod, back out throttle lever adjusting screw "C" Fig. 9, and close throttle tight. Using gauge T-109-26, loosen nut "B" Fig. 9, and move pin until it seats in notch of gauge. Tighten nut securely. Remove gauge and install metering rod, disc, and connect spring through hole in metering rod.

amount of fuel to be discharged from the Jet No. 33. If the valve cannot be cleaned with compressed air, it must be replaced.

If the Accelerator Pump Discharge Valve (Disc check) No. 30 leaks, air will be drawn into the pump cylinder on the upstroke of the plunger. This gives an insufficient charge of fuel into the throat of the carburetor upon acceleration causing a flat spot. If the valve cannot be cleaned with compressed air so that it works properly, it must be replaced.

If the Accelerating Pump Arm Spring No. 35 is weak or damaged, it will cause poor acceleration.

If the hole in the Accelerating Pump Jet No. 33 is too large, the accelerating charge will be allowed to pass too fast and will make the mixture too rich. An enlarged jet must be replaced. A jet loose on its seat gives the same effect. A clogged jet will result in a stumble on acceleration.

To adjust the pump stroke, the pump gauge T-109-117S should be used. First back out the throttle adjusting screw "C", Fig. 9, so that it does not touch the casting. In ganging the pump stroke, place the gauge on top of the bowl cover, open the throttle wide then measure to the top of the pump rod. Close throttle tight and measure again. The difference should be $^{17}\!\!/_{64}$". To adjust the stroke, bend the throttle connector rod at "A" Fig. 10. ALWAYS SET THE PUMP BEFORE SETTING THE METERING ROD. If set afterwards the metering rod will be thrown out of adjustment.

Throttle Connector Rod and Throttle Shaft Arm Assembly may be worn, and allow the throttle valve to be opened by the accelerator pedal before the pump jet begins to discharge gasoline, resulting in a flat spot. Replace all worn parts, because the operation of the metering rod is also affected.

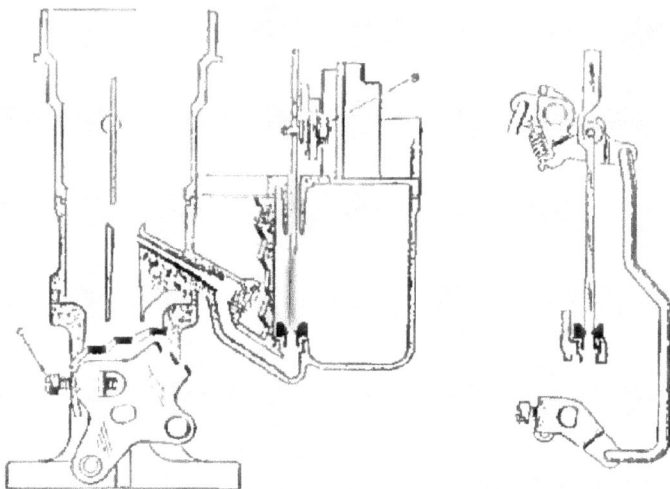

FIG. 9—METERING ROD GAUGING

Accelerating Pump Circuit

If the pump plunger is worn, sticks, or the spring under the leather has lost its tension, replace the plunger assembly No. 5, Fig. 1.

If the Accelerator Pump Intake Valve (ball check) No. 29, Fig. 6 leaks, part of the pump discharge will be forced back through the valve into the float bowl, thereby causing an insufficient

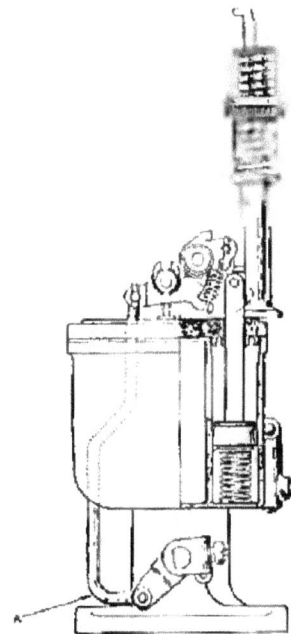

FIG. 10—PUMP TRAVEL GAUGING

Choke Circuit

The choke connector link No. 41, Fig. 7 connects the choke lever and the throttle lever and causes the throttle to be opened slightly when the choke valve is closed, thus insuring quick starting and freedom from stalling during the warm up period.

Carburetor information reprinted by permission of Carter Carburetor Corp., St. Louis, Mo., U. S. A. Holders of the copyright.

Accelerator and Linkage

The accelerator linkage, Fig. 11 is properly adjusted when vehicle leaves the factory, however, in time component parts will become worn and require adjusting to maintain a smooth even control of engine speed.

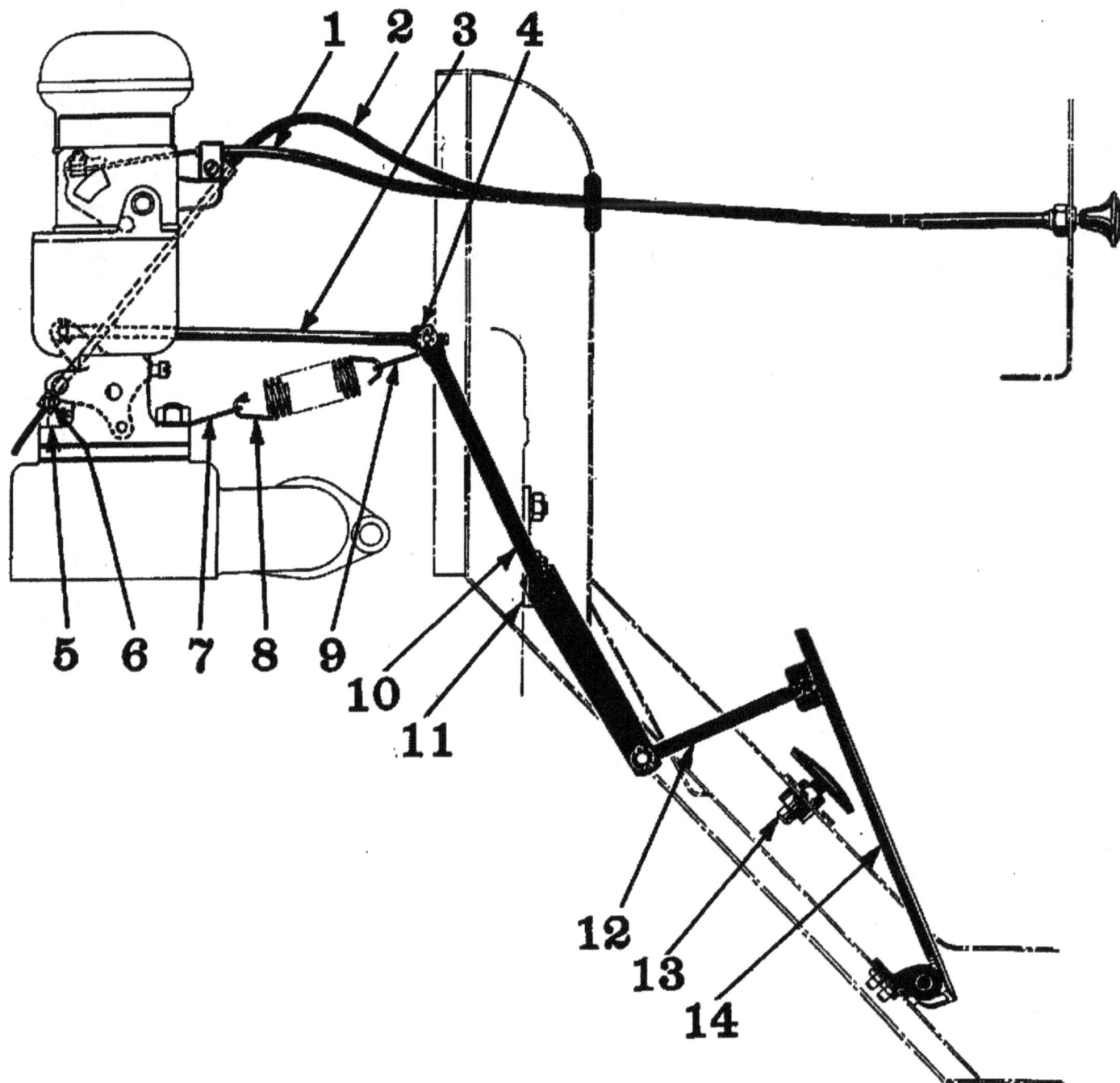

FIG. 11—ACCELERATOR, THROTTLE and CHOKE CONTROL

No.	Willys Part No.	Ford Part No.	Name		No.	Willys Part No.	Ford Part No.	Name
1	A-1301	GPW-9700	Choke Control Assembly		8	633011	GPW-9799	Accelerator Spring
2	A-5106	GPW-9775	Throttle Control Assembly		9	639610	GPW-9745	Accelerator Retracting Spring Clip
3	A-1175	GPW-9742	Throttle Rod		10	A-1243	GPW-9730	Accelerator Cross Shaft and Lever
4	633013	GPW-9752	Throttle Rod Adjusting Block		11	639607	GPW-9728	Accelerator Cross Shaft Bracket
5	50922	33800-S2	Carburetor to Inlet Manifold Stud Nut		12	A-1174	GPW-9727	Accelerator Connector Link
6	372438	GPW-11474	Throttle Wire Stop		13	A-1225	GPW-9716	Accelerator Foot Rest Assembly
7	A-1173	GPW-9751	Accelerator Retracting Spring Clip		14	A-1083	GPW-9735	Accelerator Treadle Assembly

Adjust the length of throttle rod No. 3 so when carburetor throttle valve is wide open, the accelerator treadle No. 14 will not strike the toe board. Tighten lock nut on adjusting block No. 4.

Fuel Pump

The Fuel Pump, Fig. 12 delivers a pressure of 1½ to 2½ lbs. maximum pressure at 1800 r.p.m. 10" above pump outlet.

The rotation of camshaft eccentric actuates

forcing fuel from chamber "F" through outlet valve "K" and out through "L" to the carburetor.

When the carburetor bowl is full, the float in the carburetor will shut off the needle valve, thus creating a pressure in pump chamber "F". This pressure will hold diaphragm assembly "D" downward against spring pressure "E" where it will remain inoperative until the carburetor requires further fuel and the needle valve opens. Spring "M" is merely for the

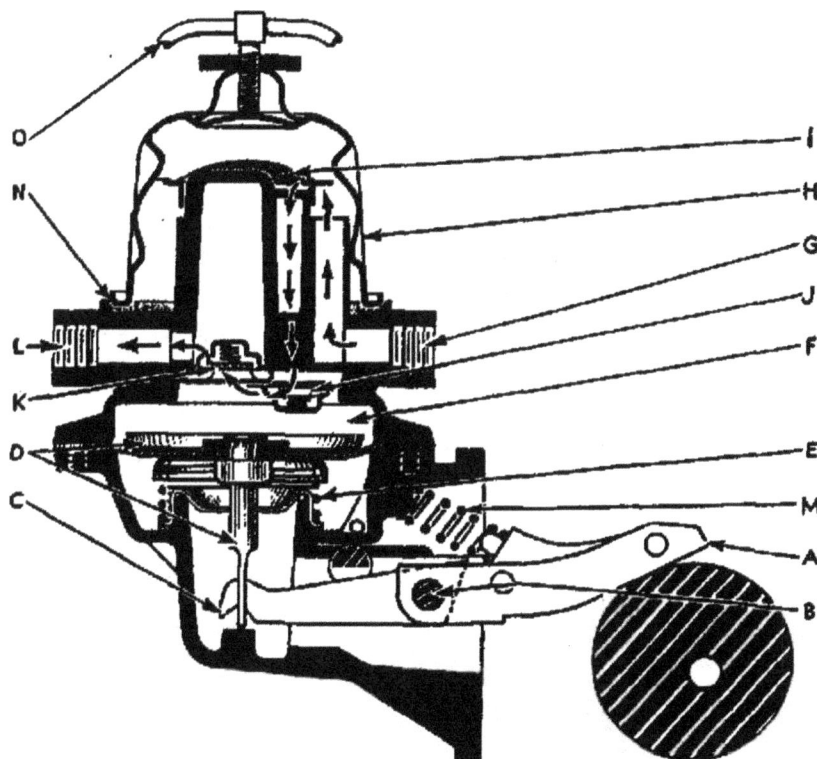

FIG. 12—FUEL PUMP

No.	Willys Part No.	Ford Part No.	Name		No.	Willys Part No.	Ford Part No.	Name
A	115641	GPW-9399	Fuel Pump Rocker Arm Assembly		H	A-1494	GPW-9355	Fuel Pump Bowl
B	A-1046	GPW-9378	Fuel Pump Rocker Arm Pin		I	115654	GPW-9365	Fuel Pump Filtering Screen Assembly
C	115880	GPW-9381	Fuel Pump Rocker Arm Link		J	115651	GPW-9352	Fuel Pump Inlet Valve Assembly
D	115644	GPW-9398	Fuel Pump Diaphragm and Pull Rod Assembly		K	115651	GPW-9352	Fuel Pump Outlet Valve Assembly
					L			Fuel Outlet
E	115865	GPW-9396	Fuel Pump Diaphragm Spring		M	115643	GPW-9380	Fuel Pump Rocker Arm Spring
F			Pump Chamber		N	115656	GPW-9364	Fuel Pump Bowl Gasket
G			Fuel Inlet		O	115657	GPW-9387	Strainer Ball Assembly

rocker arm "A" about ¼", pivoted at "B" which pulls link "C" and diaphragm assembly "D" downward against spring pressure "E" which creates a vacuum in pump chamber "F".

On the suction stroke of the pump, fuel from the tank enters inlet "G" into sediment bowl "H" and passes through strainer "I" and inlet valve "J" into pump chamber "F". On the return stroke spring pressure "E" pushes diaphragm "D" upward, purpose of keeping the rocker arm in constant contact with eccentric.

A lever and spring located on rear side of the fuel pump body is used for priming of the carburetor.

Moving the lever up and down operates the fuel pump diaphragm manually and pumps the fuel from the tank, filling the filter and carburetor bowl.

This provides a means of filling the fuel lines and units without using the starting motor, which would create unnecessary drain on battery.

The lever should be placed in the downward position to make pump function normally.

Diaphragm "D" is composed of several layers of specially treated cloth, which is impervious to fuel.

The fuel pump has a large reservoir and surge chamber. The filter bowl is clamped to the cover assembly, making it a simple matter to clean any sediment from the fuel pump. The inlet and outlet valve assemblies are interchangeable, and each assembly is a self-contained unit made up of a valve cage, a fibre valve, and a valve spring. Both valve assemblies are held in place by a valve retainer, permitting easy and speedy removal of the assemblies.

To disassemble fuel pump release thumb nut holding clamp of filter bowl, "H" and remove bowl. Remove strainer "I" from center tower, remove cork gasket, remove the six screws holding the cover flange to the pump body. Scratch a line across the two castings so that on assembly these two parts can be correctly assembled. Lift off top cover which brings into view the diaphragm assembly "D". Remove spring "M" that holds the rocker arm "A" against the camshaft eccentric.

To unhook the diaphragm pull rod "D" from the rocker arm link "C" press down and away from the rocker arm side. Remove oil seal and washer.

Remove the two screws holding inlet and outlet valve retainer.

Wash all parts thoroughly in cleaning solution and make the assembly as follows:

Install oil seal, (rubber cup) on body, then steel washer and spring that fits on the diaphragm assembly, holding the rocker arm "A" down, press the diaphragm assembly "D" down, tilt the diaphragm away from rocker arm and hook into place. Install inlet valve assembly "J" with new gasket. The inlet valve is installed in the body with the spring facing down. Install the outlet valve assembly "K" with the outlet valve spring up. Install valve retaining plate and two screws. Assemble upper and lower castings so that scratch marks line up. Install the six retaining screws and tighten them evenly. Install cam lever spring "M". Install new bowl gasket, filter screen "I" and bowl "H", tightening into place with thumb nut.

Fuel Filter

The fuel filter, Fig. 13 is of the multiple disc type bolted to the right side of the cowl and located in the fuel line between the fuel tank and the fuel pump. This is an added precaution against water or dirt reaching the carburetor. Drain the filter every few days to remove accumulated dirt and water. Be sure to tighten drain plug securely after draining.

To clean filter, remove the cover cap screw, No. 2, and remove the bowl, No. 10. Remove the filter unit, No. 8, and wash in any suitable cleaning solution. Blow out lightly with an air hose and clean out filter bowl. When replacing the unit be sure spring, No. 9, is placed over center post at the bottom of the bowl. Be sure that the gasket, No. 7, at top of filter unit and the bowl gasket, No. 6, are in good condition and in place. If the bowl gasket leaks, air will enter the fuel supply. Check for any fuel leaks with the engine stopped. Clean the filter unit every 500 operating hours or more often under severe conditions.

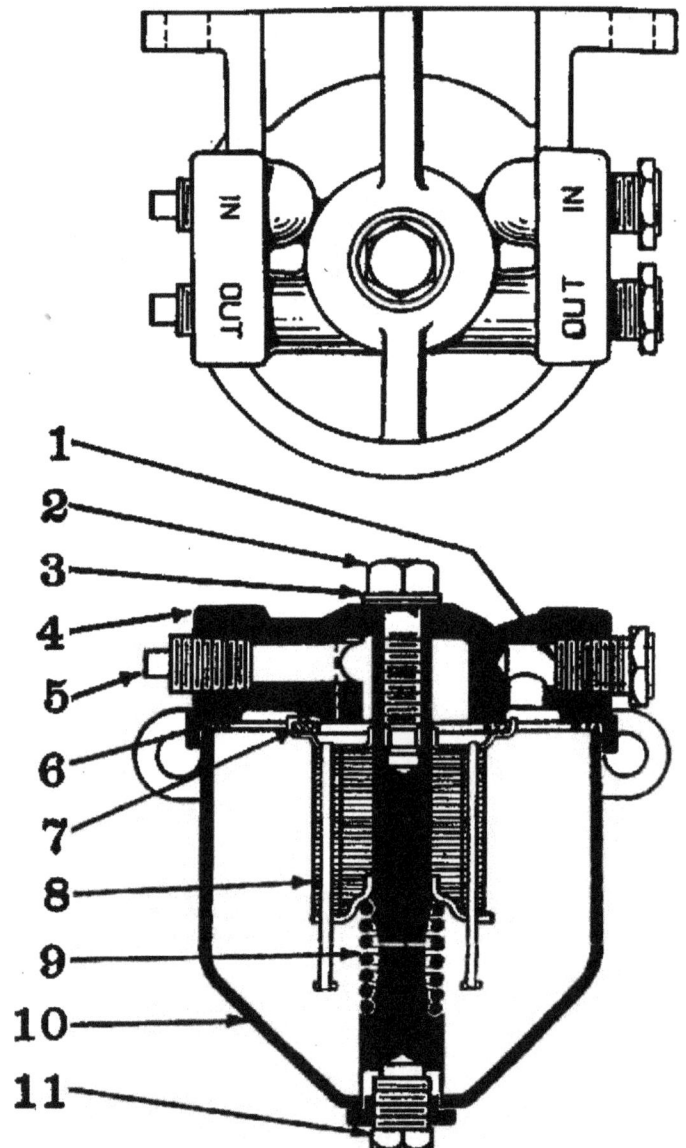

FIG. 13—FUEL FILTER

No.	Willys Part No.	Ford Part No.	Name
1	A-1255	GPW-9154	Reducing Pipe Bushing—¼ x ⅛ Pipe
2	A-1256	GPW-9183	Strainer Cover Cap Screw
3	A-1257	GPW-9184	Strainer Cover Cap Screw Gasket
4	A-1258	GPW-9149	Strainer Cover
5	5138	353055-S	Pipe Plug—¼"
6	A-1259	GPW-9160	Strainer Bowl Gasket
7	A-1260	GPW-9186	Strainer Unit Gasket
8	A-1261	GPW-9140	Strainer Unit Assembly
9	A-1262	GPW-9182	Strainer Unit Spring
10	A-1263	GPW-9162	Strainer Bowl and Center Stud
11	A-1264	GPW-9185	Strainer Drain Plug

Air Cleaner

The air which is taken into the carburetor, to mix with the fuel, is thoroughly cleaned when passed through the oil bath air cleaner Fig. 14. The cleaner is mounted on the right hand side of the dash and can be readily removed.

To clean the filter, loosen the hose clamp and two wing nuts at the center of the dash; remove the two wing nuts on the right side and remove air cleaner assembly from vehicle. Next, unfasten both clamps holding the oil cup. Unscrew element wing nut and pull out filter unit.

Wash the filter in cleaning solution by slushing it back and forth, then with an air hose dry off the unit. Do not reoil. Fill the oil cup to the indicated level. (See Capacity Chart, Page 3 and Lubrication Chart, Page 12). To assemble, reverse the dismantling procedure.

Fuel Tank Cap

The fuel tank cap is of the pressure type which keeps up to 1½ to 2 lbs. vapor pressure on the fuel. This reduces leakage and fire hazard, also eliminates depreciation of fuel qualities by evaporation.

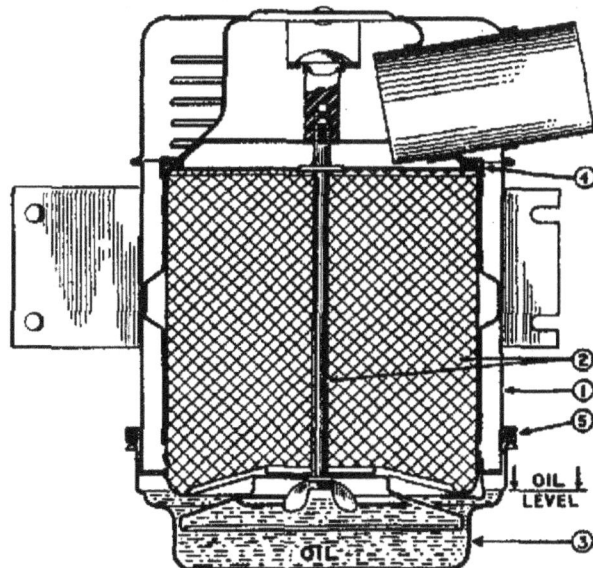

FIG. 14—AIR CLEANER (HEAVY DUTY—OIL TYPE)

No.	Willys Part No.	Ford Part No.	Name
1	A-5629	GPW-9609	Air Cleaner Body
2	A-5630	GPW-9617	Cleaner Element & Wing Bolt
3	A-5631	GPW-9658	Cleaner Cup (Oil)
4	A-5632	GPW-9621	Body Gasket (Upper)
5	A-5633	GPW-9623	Oil Cup Gasket (Lower)

FUEL SYSTEM TROUBLES AND REMEDIES

SYMPTOM	PROBABLE REMEDY

Excessive Fuel Consumption:

Tires Improperly Inflated	Inflate to 30 lbs.
Brakes Drag	Adjust
Engine Operates too Cold	Check Thermostat
Heat Control Valve Inoperative	Check Thermostatic Spring
Leak in Fuel Line	Check All Connections
Carburetor Float Level High	See Carburetor Section
Accelerator Pump Not Properly Adjusted	Adjust
Leaky Fuel Pump Diaphragm	Replace
Loose Engine Mountings (High carburetor fuel level)	Tighten
Ignition Timing Slow or Spark Advance Stuck	See Distributor Section
Low Compression	Check Valve Tappet Clearance
Air Cleaner Dirty	Remove and Clean

Engine Hesitates on Acceleration:

Accelerator Pump does not Function Properly	Replace Piston and Rod or Adjust
Carburetor Float Level	Adjust
Spark Plugs	Replace or Clean and Adjust
Low Compression	Check Valves
Distributor Points—Dirty or Pitted	Replace
Weak Condenser or Coil	Replace
Carburetor Jets Restricted	Remove and Clean
Excessive Engine Heat	See Engine Section

Engine Stalls—Won't Idle:

Improper Condition of Carburetor	See Carburetor Section
Low Speed Jet Restricted	Remove and Clean
Dirty Fuel Sediment Bowl Screen	Remove and Clean
Air Cleaner Dirty	Remove and Clean
Leaky Manifold or Gasket	Replace
Fuel Pump Diaphragm Porous	Replace
Loose Carburetor	Tighten Flange Nuts
Water in Fuel	Drain System
Improper Ignition	See Distributor Section
Spark Plugs	Clean and Adjust
Valves Sticking	Grind Valves

FUEL SYSTEM SPECIFICATIONS

Carburetor:

Make.................................Carter

Model.........................W.O.-539 S

Flange..................................1"

Primary Venturi...................11/32" I.D.

Main Venturi.......................1" I.D.

Float Setting...........................3/8"

Fuel Intake..... Square Vertical Spring Loaded Needle No. 53 Drill Size in Needle seat.

Fuel Line Connection.....1/8" pipe thread—3/16" inverted flared tube elbow.

Low Speed Jet Tube
Jet Size......................No. 71 Drill
Idle Well Jet.................No. 61 Drill

Idle Screw Seat.................No. 46 Drill

Main Nozzle Discharge Jet.......096" Diameter

Metering Rod...................No. 75-547
Jet Size....................070" Diameter
Setting (Use gauge No. T-109-26).....2.718"

Accelerating Pump
Discharge Jet..................No. 73 Drill
Intake Ball Check.............No. 40 Drill
Discharge Disc Check..........No. 40 Drill
Relief Passage to Outside.......No. 42 Drill
Adjustment (Use gauge T-109-117 S)....17/64"

Air Cleaner:

Make..............................Oakes
Model............................613300
Type............................Oil Bath
Oil Capacity.......See Capacity Chart, Page 3

Fuel Pump:

Make................................AC
Model...............................AF
Type............................Camshaft
Pressure..11/2 to 21/2 lbs. at 16" above outlet @ 1800 R.P.M.

Fuel Tank:

Make................................Own
Capacity.........See Capacity Chart, Page 3
Location...............Under Driver's Seat
Filler Cap....................AC No. 850018

Fuel Filter:

Make...............................AC
Model...............................T-2
Type...............................Disc
Mounting...................Mounted on Dash

EXHAUST SYSTEM

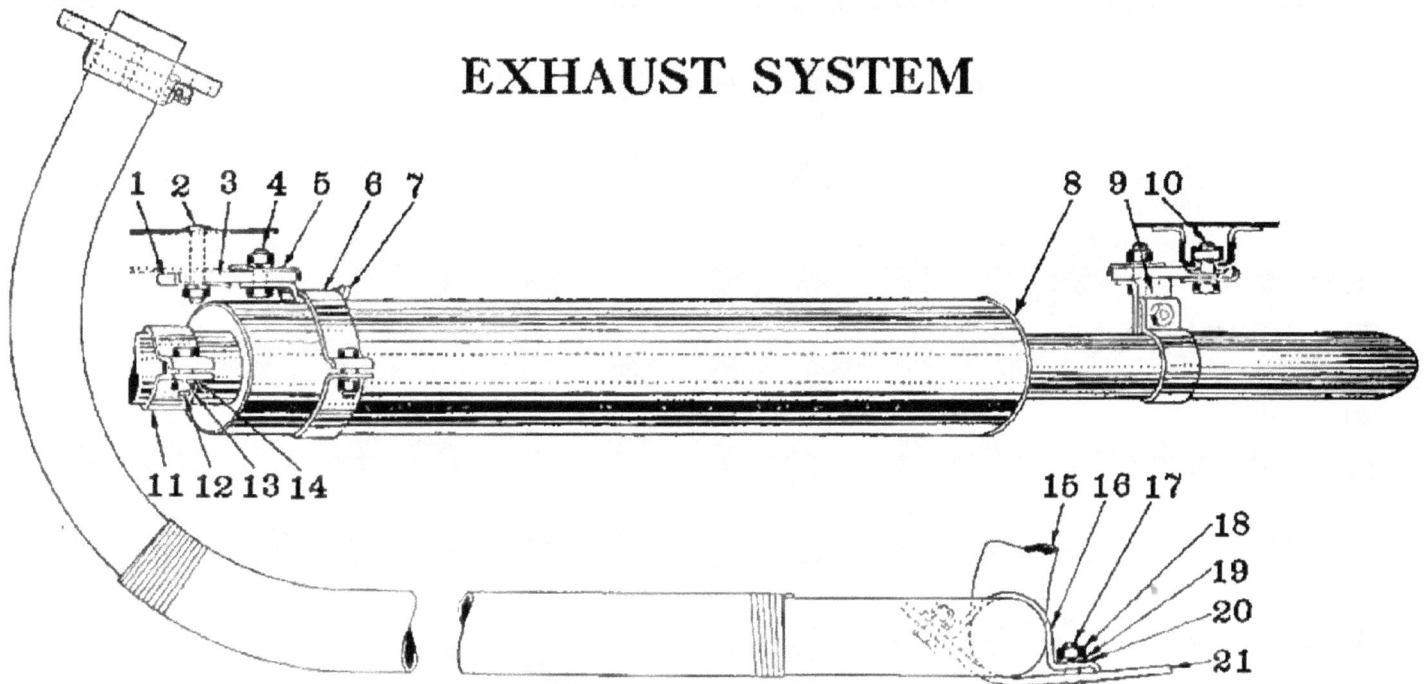

FIG. 1—EXHAUST SYSTEM

No.	Willys Part No.	Ford Part No.	Name
1	638058	GPW-5274	Muffler Support Insulator Plate
2	52372	23498-S	Body Sill to Muffler Support Bolt
3	A-658	GPW-5283	Muffler Support Insulator
4	50929	20367-S7	Muffler Support Screw
5	638058	GPW-5274	Muffler Support Insulator Plate
6	A-5753		Muffler Support Clamp
7	A-657	GPW-5262	Muffler Support Strap
8	A-6118		Muffler Assembly
9	A-6119		Muffler Tail Pipe Clamp
10	50929	24367-S	Muffler Support Screw
11	536004	GPW-5270	Exhaust Pipe to Muffler Clamp
12	5922	24407-S	Exhaust Pipe to Muffler Clamp Screw
13	5010	33798-S	Exhaust Pipe to Muffler Clamp Screw Nut
14	51833	34806-S	Exhaust Pipe to Muffler Clamp Screw Nut Lockwasher
15	A-1296	GPW-5246	Exhaust Pipe Assembly
16	A-1300	GPW-5251	Exhaust Pipe Extension Clamp (to Skid Plate)

No.	Willys Part No.	Ford Part No.	Name
17	52983	23393-S	Exhaust Pipe Extension to Skid Plate Bolt
18	6107	33797-S	Exhaust Pipe Extension to Skid Plate Bolt Nut
19	51833	34806-S	Exhaust Pipe Extension to Skid Plate Bolt Lockwasher
20	52274	34746-S2	Exhaust Pipe Extension to Skid Plate Bolt Plain Washer
21	A-1253	GPW-5291-B	Underframe Skid Plate

Seamless exhaust pipe and flexible metal tubing connect the Manifold to the Muffler.

The Muffler, No. 8, Fig. 1 is designed for straight through exhaust to minimize back pressure. The Muffler is attached to the under side of the vehicle on the right hand side by means of support straps and flexible Insulators.

The exhaust pipe slides into the nipple on the front end of the Muffler and is held in place by means of a clamp, No. 11.

Exhaust and Intake Manifold

The exhaust and intake Manifolds, make a unit in which the hot exhaust gases are thermostatically controlled, and directed around the intake Manifold to assist in vaporizing the fuel when engine is cold, thereby aiding in warming up the engine and reducing oil dilution. It also minimizes the use of the carburetor choke control and results in proper temperature of the incoming gases under all operating conditions.

When the engine is cold, the counterweight lever No. 6, Fig. 2, closes the valve and directs the hot exhaust gases against the intake Manifold. As the engine warms up, the thermostatic (or Bimetal) spring No. 7 expands and opens the valve directing the exhaust gases into the exhaust pipe.

When assembling the Manifolds to the cylinder block, new gaskets should be installed, and the nuts drawn up evenly until they are all tight to avoid gas leakage. Torque wrench reading, 31-35 ft. lbs.

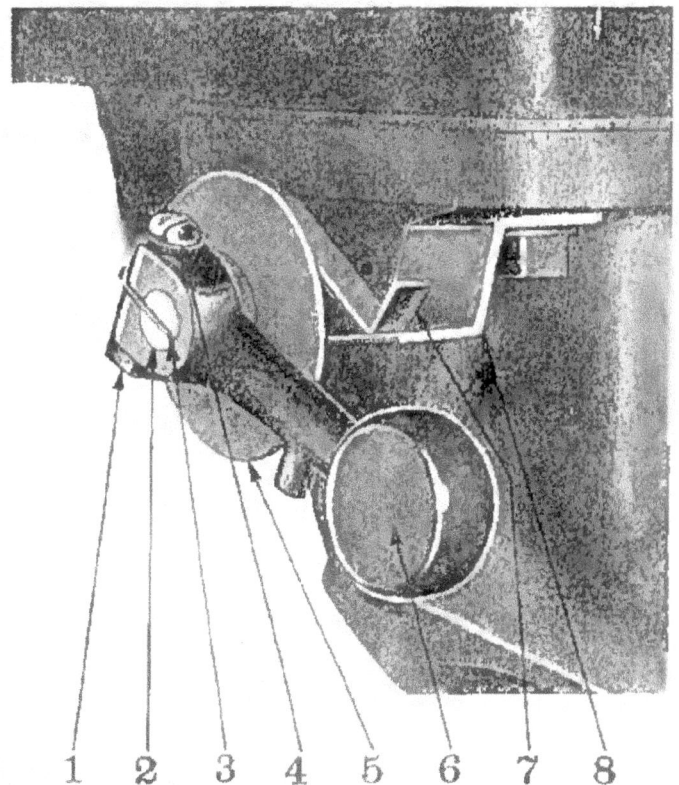

FIG. 2—HEAT CONTROL VALVE

No.	Willys Part No.	Ford Part No.	Name
1	6353	355836-S7	Heat Control Valve Lever Clamp Screw Nut
2	637206	GPW-9456	Heat Control Valve Shaft
3	637211	GPW-9463	Heat Control Valve Lever Key
4	5272	355160-S	Heat Control Valve Lever Clamp Screw
5	637209	GPW-9484	Heat Control Valve Bi-Metal Spring Washer
6	637210	GPW-9458	Heat Control Valve Counterweight Lever
7	637208	GPW-9467	Heat Control Valve Bi-Metal Spring
8	639743	GPW-9463	Heat Control Valve Bi-Metal Spring Stop

COOLING SYSTEM

The satisfactory performance of an engine is controlled to a great extent by the proper operation of the cooling system. The engine block is full length water jacketed which avoids distortion of the cylinder walls. Directed cooling and large water holes, properly placed in cylinder head gasket, cause more water to flow past the valve seats (which is the hottest part of the block) and carry the heat away from the valves, giving positive cooling of valves and seats.

To quickly warm up the engine and hold the cooling fluid to the maximum efficient temperature, there is a thermostat installed in the water outlet on the cylinder head.

Radiator

The radiator is designed to cool the water under all operating conditions, however, the radiator core must be kept free from corrosion and scale in addition to the maintenance of other cooling units to obtain satisfactory service.

At least every 20,000 miles remove the radiator and clean it inside and out in a cleaning solution. At the same time examine core for leaks or damaged cells.

After radiator and cooling system have been cleaned and flushed out, it is advisable to use a corrosion preventative. Rust and scale may eventually clog up water passages in both the radiator and water jacket of the engine unless a rust inhibitor is used. This condition is aggravated in some localities by the water available.

Emergency repairs in case of puncture by bullet or shrapnel; if a tube is not completely severed, cut it or break it off with a pair of pliers. With pliers strip fins from tube above and below break for ½" or necessary distance to enable bending of the tube around itself and flatten, both above and below the break thereby stopping the flow of water.

Radiator Filler Cap

The cap is of the pressure type which prevents evaporation and loss of cooling solution. A pressure up to 3¼ to 4¼ pounds makes the engine more efficient by running at a slightly higher temperature. Vacuum in the radiator is relieved by a vacuum valve opening at ½ to 1 pound vacuum.

Draining Cooling System

To drain the cooling system open the drain cock located at the lower left hand corner of the radiator, just under the water outlet, also the drain cock at right front lower corner of cylinder block.

Remove radiator cap to break any vacuum and thoroughly drain system.

Filling the Cooling System

Close drain cocks in the cylinder block and radiator. Fill the radiator with clean water or during cold weather with an anti-freeze solution. Do not overfill the radiator while anti-freeze solution is being used, because the solution expands when heated and an appreciable amount of liquid would be lost through the overflow. The solution should be 1" from the bottom of the filler neck.

Should water be lost from the cooling system and the engine overheats, do not add water immediately but allow the engine to cool, then add water slowly while the engine is running.

If cold water is poured into the radiator while the engine is overheated, there is danger of cracking the cylinder block and head.

Thermostat

The cooling system is designed to provide adequate cooling under the most adverse conditions; however, it is necessary to employ some device to prevent overcooling during normal operations. This is accomplished by use of a thermostat, (No. 39, Fig. 1, "Engine" Section) which is located in the water outlet on top of the cylinder

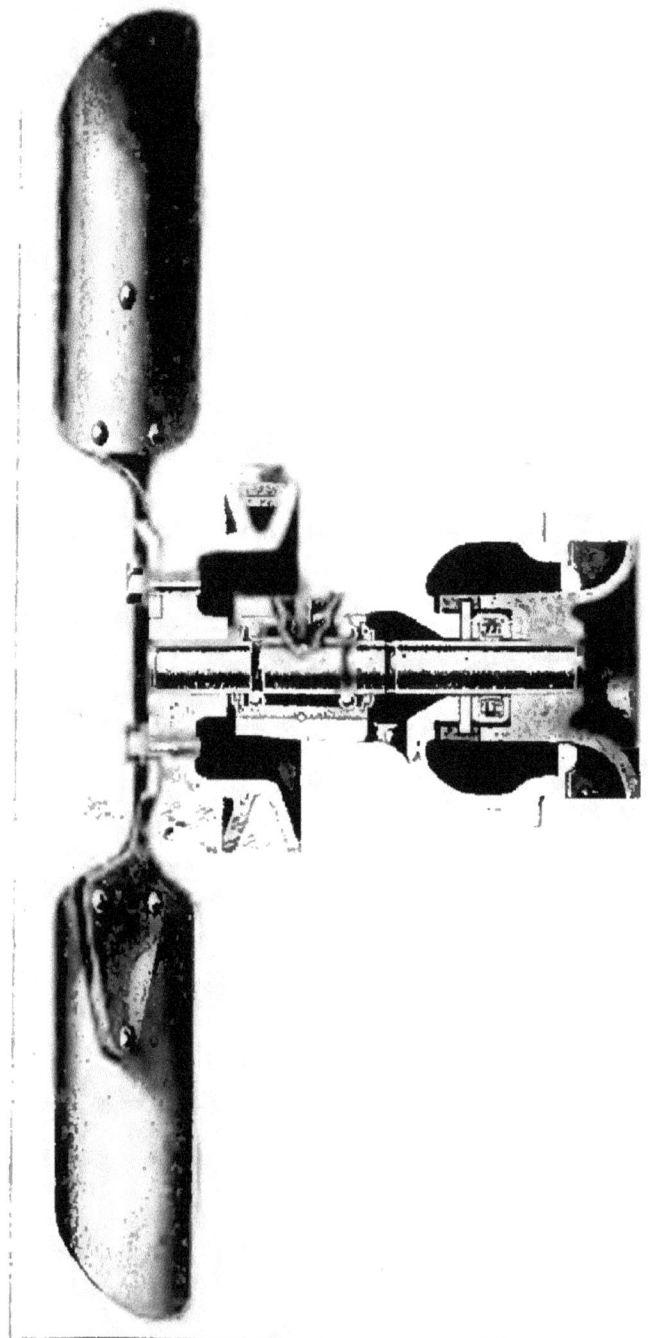

FIG. 1—WATER PUMP AND FAN ASSEMBLY

head. The thermostat opening is set by the manufacturer and cannot be altered. The thermostat opens at a temperature of 145° to 155° Fahr. To test thermostat, heat sufficient water to 170° Fahr. and submerge thermostat. The valve should open to the limit at this temperature. If valve fails to open, a new thermostat will be required.

Heat Indicator

The heat indicator is of the Bourdon type and is connected to a bulb in the engine block by means of a capillary tube.

If the unit becomes inoperative, it should be replaced as it is not practical to either repair or adjust this unit.

Water Pump

The water pump, Fig. 1 is a centrifugal impeller type of large capacity to circulate the water in the entire cooling system.

The double row ball bearing is integral with the shaft, No. 2, Fig. 2 and is packed with a special high melting point grease at the time of manufacture so it requires no lubrication. The ends of the bearings are sealed to retain the lubricant and prevent dust and dirt from entering.

The bearing is retained in the housing by a retaining wire No. 4, which snaps between the bearing and the water pump body. The seal washer No. 5 has four lugs which fit into the slots in the end of the impeller No. 8. One side of the seal washer bears against the ground surface of the pump housing and the other against the seal No. 7. The rubber seal bears against the machined surface on the inside of the impeller. The seal maintains a constant pressure against the seal washer and impeller assuring positive seal. The drain hole in the bottom of housing prevents any water seepage past the seal washer No. 5 entering the bearing.

The impeller and fan pulley are pressed on to the straight shaft under 2500 pounds pressure.

Dismantling of Water Pump

Remove fan belt and fan assembly and then the water pump from the engine.

Remove bearing retaining wire No. 4, Fig. 2.

Place water pump body on arbor press face plate and press water pump shaft through impeller No. 8, and pump body No. 3.

Remove the seal washer No. 5 and seal No. 7.

Place pump shaft No. 2 and fan pulley No. 1 on press so that the bearing will clear in the opening and press shaft from fan pulley.

To reassemble the water pump; install the long end of the shaft No. 2, in the pump body, No. 3 from the front end, until the outer end of bearing is flush with the front end of the pump body.

Dip seal No. 7 and seal washer No. 5 in brake fluid and install in the impeller No. 8. Place the impeller on an arbor press and press the long end of shaft into the impeller, until the end of the shaft is flush with the impeller.

Support assembly on impeller end of shaft and press the fan pulley on to shaft so the end of shaft is flush with the face of fan pulley—move the shaft in the pump body so grooves in the bearing and pump body line up and install bearing retaining wire No. 4.

Anti-Freeze Solution

Where air temperatures require, it is necessary to protect the cooling system with some type of anti-freeze solution so as to prevent damage resulting from freezing.

When alcohol is used as an anti-freeze solution care must be taken not to spill any of the solution on the finished portions of the vehicle; if so, it should be washed off immediately with a good supply of cold water, without wiping or rubbing.

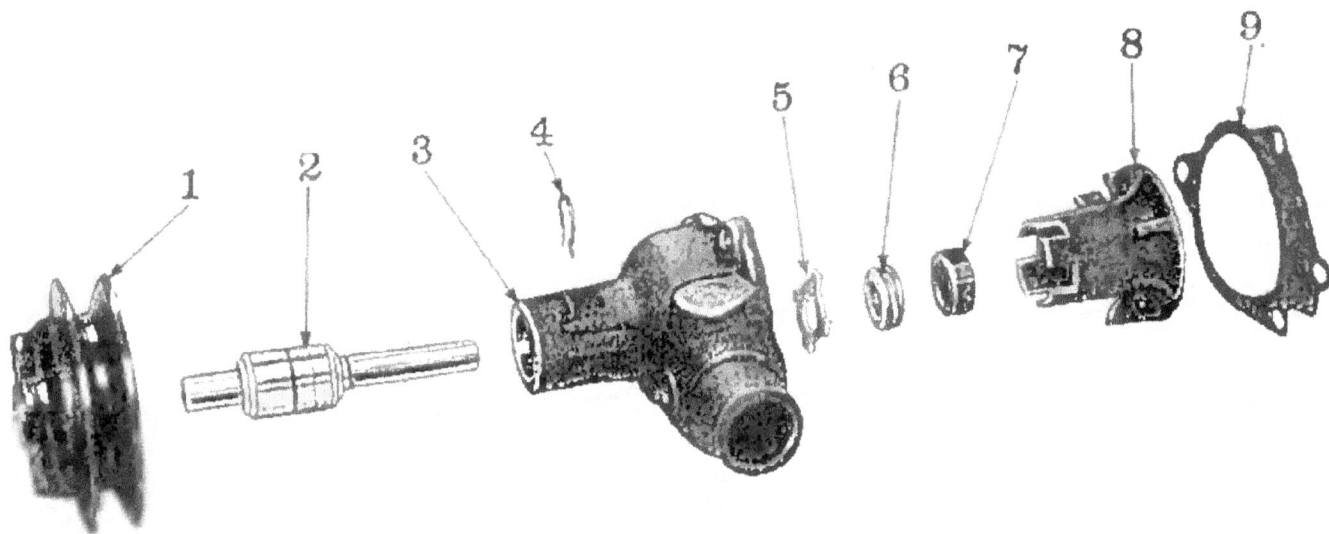

FIG. 2—WATER PUMP ASSEMBLY

No.	Willys Part No.	Ford Part No.	Name	No.	Willys Part No.	Ford Part No.	Name
1	636299	GPW-8509-A	Fan and Water Pump Pulley	5	639994	GPW-8557	Water Pump Seal Washer
2	636297	GPW-8530	Water Pump Bearing and Shaft Assembly	6-7	639663	GPW-8524	Water Pump Seal Assembly
3	637052	GPW-8505	Water Pump Body	8	639993	GPW-8512	Water Pump Impeller
4	636298	GPW-8576	Water Pump Bearing Retaining Wire	9	637053	GPW-8543	Water Pump to Cylinder Block Gasket

The distillation or evaporating point of water and alcohol is approximately 170° Fahrenheit, therefore, when the engine is operated in warm weather with alcohol solution, the solution must be checked regularly as there will be considerable loss of the alcohol through vaporization, and the freezing point raised in the solution, this might cause freezing of the solution at a sudden drop in temperature.

Ethylene glycol anti-freeze solutions have the distinct advantage of possessing a higher point of distillation than alcohol and consequently may be operated at higher temperatures without loss of the solution through evaporation.

Ethylene glycol has the further advantage that in a tight system only water is required to replace evaporation losses, however, any solution lost mechanically through leakage or foaming must be replaced by additional new solution. Under ordinary conditions Ethylene glycol solutions are not injurious to the car finish.

Rust and scale forms in every cooling system, therefore, we recommend that the cooling system be flushed out twice a year preferably before and after using anti-freeze. There are a number of flushing solutions and the instructions of the manufacturer should be closely followed when they are used.

Remove the thermostat when flushing the engine block, so the water and air pressure can get by and avoid possible damage to the thermostat.

When the cooling system is being conditioned it is good policy to tighten the cylinder head bolts to prevent the possibility of water leaking into the cylinders and lubrication oil. The radiator hoses should be inspected regularly for any indication of leakage which might be caused by loose clamps or deteriorated hose.

Fan Belt

The fan is driven by a "V" Belt. Angle of "V" -42°. Length outside 44⅛". Width maximum 1¹¹⁄₁₆".

To install Fan Belt loosen clamp bolt on slotted bracket at generator and move generator towards engine. Slide belt over crankshaft pulley, up through fan blade assembly and over fan pulley, then over generator pulley. Adjust the fan belt by bringing the generator away from the engine to a point where the fan belt can be depressed 1" midway between fan pulley and generator pulley. The drive of the fan and generator is on the sides of the "V" belt, therefore it is not necessary to have the fan belt tight which might cause excessive wear on generator and water pump bearings.

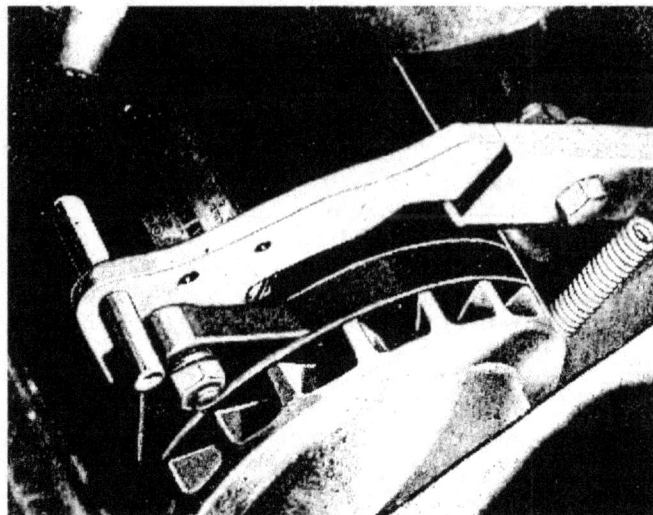

FIG. 2—GENERATOR BRACE

When there is a possibility of water being thrown over the engine by fan action in crossing streams, pull up on the handle of the generator brace, then remove the fan belt. As soon as possible the belt should be replaced, then pull out on the generator. The generator will lock in place by spring action of the brace.

COOLING TROUBLES AND REMEDIES

SYMPTOMS	PROBABLE REMEDY
Overheating	
Lack of Water	Refill Radiator
Thermostat Inoperative	Replace
Water Pump Inoperative	Overhaul or Replace
Incorrect Ignition or Valve Timing	Set Timing
Excessive Piston Blowby	Check Pistons, Rings and Cylinder Walls
Fan Belt Broken	Replace
Radiator Clogged	Reverse Flush
Air Passages in Core Clogged	Clean With Water and Air Pressure
Excessive Carbon Formation	Remove
Muffler Clogged or Bent Exhaust Pipe	Replace
Loss of Cooling Liquid	
Loose Hose Connections	Tighten
Damaged Hose	Replace
Leaking Water Pump	Replace
Leak in Radiator	Remove and Repair
Leaky Cylinder Head Gasket	Replace
Crack in Cylinder Block	Small Crack Can Be Closed with Radiator Anti-Leak
Crack in Cylinder Head	

COOLING SYSTEM SPECIFICATIONS

Cooling Capacity....See Capacity Chart, Page 3

Radiator.........................Jamestown

   Radiator Filler Cap....................A C

Fan—4 Blade—15" Dia...............Hayes

Fan Belt

   Type..............................."V"
   Length...........................44⅛"
   Width.............................1¼₆"
   Angle of Vee.........................42°

Water Pump

   Type........................Centrifugal
   Location.............Front of Cylinder Block
   Drive................................Belt
   Bearing...Permanently Sealed-Lubricated Ball

Thermostat

   Location......Water outlet Top Cylinder Head
   Starts To Open At.................145°-155°
   Fully Open............................170°

Anti-Freeze

Temp. Fahr.	Alcohol Qts	Ethylene Glycol Qts	Temp. Cent.
30°	1	1	— 1.1°
20°	2⅛	2	— 6.6°
10°	3¼	3	—12.2°
0°	4¼	3¾	—17.7°
—10°	5	4½	—23°
—20°	5½	4¾	—29°
—30°	6¾	5½	—34°
—40°	7¼	6	—40°

To convert quantities into Imperial quarts multiply by .833.

To convert quantities into Metric liters multiply by .946.

ELECTRICAL SYSTEM

The wiring diagrams, Fig. 1 and 3 show the general arrangement of all chassis electrical circuits, together with units in correct relation to position in which they will be found on the chassis.

Regular inspection of all electrical connections avoids failures in the electrical system. When tracing any one particular circuit refer to Wiring Harness, Page 59 for color of wire and tracer.

Radio Interference Suppression

The vehicle is equipped with filters in several of the electrical circuits and resistor type suppressors have been placed in the ignition high tension system, which together with bonding and shielding prevents interference with radio communication. (See Page 72.)

Filterettes, consisting of a coil in series with the line and two condensers across the line to ground, have been placed in the line from the primary of the high tension coil to the ignition switch; from ammeter to the "B" terminal of the voltage regulator and, from ammeter to the battery. Condenser type filters have been placed from the generator armature terminal to ground, and, from the regulator field terminal to ground. See detail wiring diagram in Fig. 1, Page 58.

Failure of the ignition or charging system because of open circuit filters can be checked by shorting across the terminals of the suspected filterette unit. If operation is not restored filterette may be considered to be in proper condition. These filterettes are combined in a container on the drivers side of the dash. There are two types of containers. Terminals are exposed on one type by opening cover which is held by a latch on the top, and on the other type by removing the cover. In an emergency, operation of the vehicle can be resumed by placing a shorting wire across the open circuit of the filterette.

To check filterette for short to ground, remove case of unit from dash. If short is NOT removed filterette may be considered satisfactory. If short IS removed filterette is defective. Temporary operation can be obtained by leaving case of the filterette ungrounded.

The regulator field filter (condenser type) which is mounted on the regulator itself can be checked by the same procedure as in the test for a shorted filterette.

Broken suppressors (resistor type) will affect the operation of the vehicle. Temporary operation can be obtained by removing the broken suppressor and making direct connection.

On vehicles equipped with radio suppression, the letter "S" appears on the cowl between the rear end of the hood and the windshield, 1¼" above bottom edge of hood.

Battery

The battery is a 6 Volt, 15 Plate, 116 Ampere Hour battery. It is located under the hood on a bracket attached to the right hand side rail of the frame and held solidly on the base with a battery hold down assembly over the top of the battery by two studs and wing nuts.

The battery should be checked once a week with a hydrometer and at the same time check the electrolyte level in each cell; add distilled water if necessary. Do not fail to replace filler caps and tighten securely. Battery fumes or acid coming in contact with any metal parts causes corrosion and eating away of these parts. To safeguard against this difficulty avoid overfilling.

The negative terminal of the battery (smallest post) is grounded by a cable bolted to the frame.
The engine ground cable located on the right hand side of engine, which connects the front engine support plate with the frame, is required due to the engine being mounted on rubber insulators.

If the terminal connections of this cable are loose or dirty, it will cause hard starting of the engine. Attention should be given to this ground cable during each inspection, and also at the time the engine requires a tuneup.

Fuel Gauge

The fuel gauge circuit Fig. 2, is composed of two units. The indicating unit or dash unit which is mounted in the instrument panel, and the tank unit, which is mounted in the fuel tank. These units are connected by a single wire. The circuit for this instrument passes through the ignition switch, therefore, the fuel gauge operates only when the ignition switch is on.

The dash unit is of the balanced coil type and is designed so that its operation is not affected by variations in the voltage of the electrical system.

The tank unit consists of a resistance wire wound on an insulator and a contact arm which is moved by the float arm. As the depth of the fuel in the tank varies, the contact arm is moved across the resistance wire and so varys the resistance. As this resistance is varied, there results a proportionate variation of current in the coils of the dash unit which is calibrated to accurately indicate the fuel level in the tank.

If the gauge does not register properly, first check all wire connections to be sure that they are clean and tight. Then make sure that the dash unit is grounded to the dash and that the tank unit is grounded to the fuel tank.

WILLYS MODEL "MB" ¼-TON 4 x 4 GOVERNMENT TRUCK

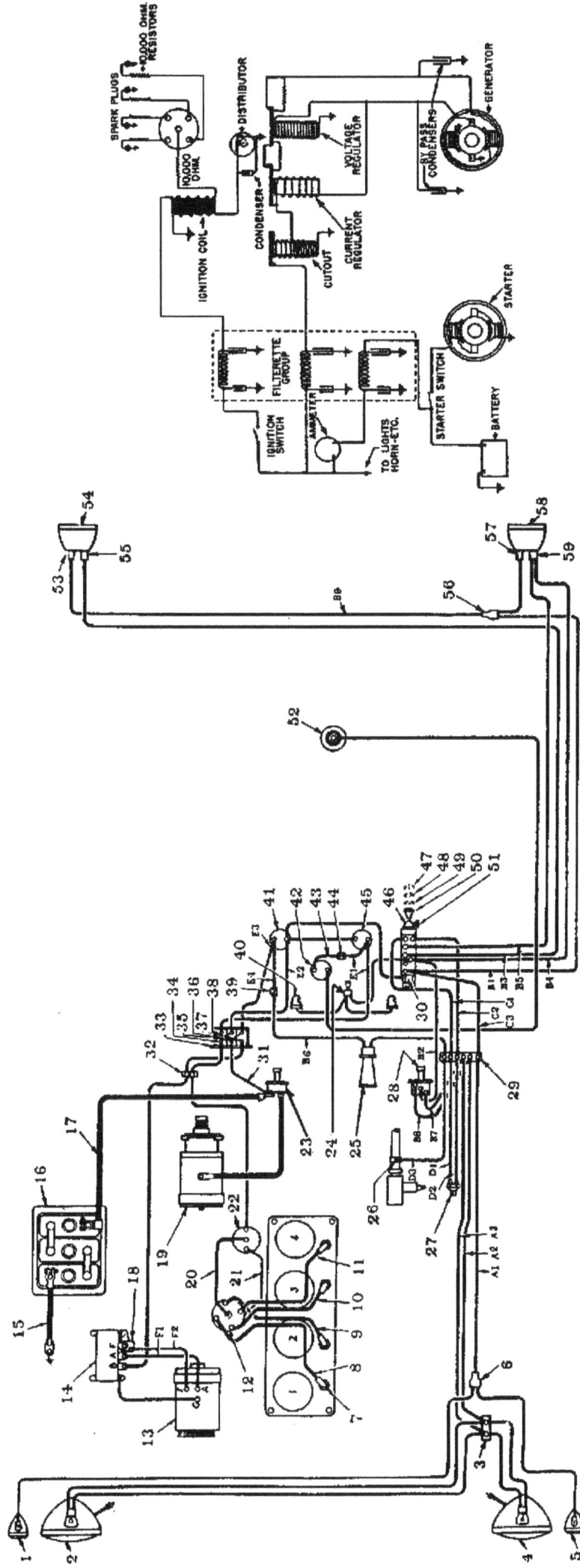

FIG. 1—WIRING DIAGRAM

No.	Ford Part No.	Willys Part No.	Name
1	GPW-13200	A-1437	Blackout Head Lamp—Right
2	GPW-13205	A-1305	Service Head Lamp—Right
3	GPW-14448-A	639599	Junction Block—2 post
4	GPW-13206	A-1304	Service Head Lamp—Left
5	GP-13700-B	A-1436	Blackout Head Lamp—Left
6	GPW-14487-A	635985	Connector
7	A-538	A-538	Spark Plug and Gasket
8	GPW-12405	A-1412	Spark Plug Cable No. 1
9	GPW-12287	A-1414	Spark Plug Cable No. 2
10	GPW-12284	A-1415	Spark Plug Cable No. 3
11	GPW-12283	A-1418	Spark Plug Cable No. 4
12	GPW-12288	A-1244	Ignition Distributor
13	GPW-12100	A-5992	Generator Assembly
14	GPW-10505	A-1409	Voltage Regulator and Circuit Breaker
15	GPW-14301	A-1320	Battery Ground Strap
16	11AS-10655	A-1238	Battery
17	GPW-14300	A-1402	Battery Positive Cable to Starter Switch
18	GPW-18026-A	A-1287	Radio Filter Unit (Generator Regulator)
19	GPW-11001	A-1245	Starting Motor
20	GPW-12298-B	A-1420	Ignition Coil Secondary Cable
21	GPW-14321	A-5063	Ignition Coil Primary Cable

No.	Ford Part No.	Willys Part No.	Name
22		A-1527	Ignition Coil
23		A-6181	Starter Switch
24	GPW-13740	A-1333	Instrument Lamp Switch
25	GPW-13802	A-1312	Horn
26	GPW-13836	A-302	Horn Button Contact (Steering Gear)
27	11A-13480	A-1271	Stop Lamp Foot Switch
28	GPW-13532	638979	Junction Block
29	GPW-14448-B	A-1490	Thermal Circuit Breaker—30 Amp.
30	GPW-12250-C	A-1349	Filter (Reg. Bat. Amm.) to Starter Switch Cable
31	GPW-14457	A-5078	Junction Block
32	639599	A-5980	Filter Group and Bracket Assembly
33	GPW-14448-A		Filter Terminal to Coil Primary
34			Filter Terminal to Battery
35			Filter Terminal to Regulator (B)
36			Filter Terminals to Ammeter
37			Filter Terminal to Ignition Switch
38			Circuit Breaker, 15 Amp.—Horn
39	GPW-12250-B	A-1734	Instrument Lamp Socket and Cable
40	GPW-13710	A-1411	

No.	Ford Part No.	Willys Part No.	Name
41	GPW-10860	A-5231	Ammeter
42	GPW-9280	A-1288	Gas Gauge Dash Unit
43	GPW-14416	A-5080	Gas Gauge Circuit Breaker to Gas Gauge
44	GPW-12250-A	A-1733	Dash Unit Cable
45		A-2517	Circuit Breaker, 5 Amp, Gas Gauge
46	11TS-11654	A-1332	Ignition Switch
47			Blackout Lighting Switch
48			Service Stop Light Position
49			Service Position
50			Blackout Position Off Position
51			Service Position Lock
52	GPW-9275	A-1292	Fuel Gauge—Tank Unit
53	GP-13404-B2	A-1065	Blackout Tail Light
54			Tail and Stop Lamp—Right
55	GPW-14487-A	635985	Blackout Stop Light
56			Connector
57	GPW-14487-A		Blackout Tail Light
58	GP-13405-B2	A-1064	Tail and Stop Lamp—Left
59			Service Tail and Service Stop Light

WIRING HARNESS

Willys Part No.	Ford Part No.	Name
A-1665	GPW-14425	Head lamp wiring harness
		The following (3) cables are included in this harness but not supplied individually:
		A-1 Blackout head lamp junction block cable (Yellow with two black tracings)
		A-2 Head lamp junction block to junction block cable (Upper beam) (Red, three white tracings)
		A-3 Head lamp junction block to junction block cable (Lower beam) (Black, white tracings)
A-5048	GPW-14401-B	Body wiring harness—Long
		The following (9) cables are included in this harness but not supplied individually:
		B-1 Light switch (Terminal marked B.H.T.) to blackout tail light Conn. Cable (Yellow, two black tracings)
		B-2 Light switch (Terminal marked H.T.) to foot dimmer switch center terminal cable (Blue, three white tracings)
		B-3 Light switch (Terminal marked B.S.) to blackout stop light cable (White, two black tracings)
		B-4 Light switch (Terminal marked H.T.) to service tail light and inst. lamp switch cable (Blue, two white tracings)
		B-5 Light switch (Terminal marked S.) to service stop light cable (Black, two white tracings)
		B-6 Horn circuit breaker to horn cable (Black, two red tracings)
		B-7 Junction block to foot dimmer switch (Lower beam) Cable (Black, two white tracings)
		B-8 Junction block to foot dimmer switch (Upper beam) Cable (Red, three white tracings)
		B-9 Connector to blackout tail light cable (Yellow, two black tracings)
A-1551	GPW-14402	Body wiring harness—Left side—Short
		The following (3) cables are included in above harness but not supplied separately:
		C-1 Junction block to light switch (Stop switch) (Red, two white tracings)
		C-2 Junction block to light switch cable (Stop switch) (Green, two black tracings)
		C-3 Junction block to light switch cable (Blackout head) (Yellow, two black tracings)
A-5061		Chassis wiring harness—Left side
		The following (3) cables are included in above harness but not supplied separately:
		D-1 Stop light switch to junction block cable (Red, two white tracings)
		D-2 Stop light switch to junction block cable (Green, two black tracings)
		D-3 Steering gear horn terminal to junction block cable (Three black, two white tracings)
A-5981		Filter wiring harness
		The following (4) cables are included in this harness but not supplied individually:
		E-1 Filter (Coil ign. switch) to ignition switch to gas gauge circuit breaker (Black, two white tracings)
		E-2 Filter (Battery, ammeter) to ammeter cable (Red, three white tracings)
		E-3 Filter (Reg. battery ammeter) to ammeter cable (Black, three white tracings)
		E-4 Ammeter to horn circuit breaker cable (Black, two red tracings)
A-5074	GPW-14305	Generator to voltage regulator and filter harness
		F-1 Generator to regulator armature cable (Red, three white tracings)
		F-2 Generator to regulator field cable (Green, two black tracings)
A-719	GPW-13410	Blackout tail lamp to connector cable, left
A-5072	GPW-14458	Ignition switch to ammeter to blackout switch cable
A-5080	GPW-14416	Fuel gauge to circuit breaker cable
A-1731	GPW-14436	Headlamp ground cable
A-5070	GPW-14406	Fuel gauge (Inst. board) to fuel gauge (Tank unit) cable
A-5081	GPW-14409	Horn to junction block cable
A-5073	GPW-14459	Filter (Reg. bat.) to junction block cable
A-5078	GPW-14457	Filter (Bat.) to starter switch cable
A-5079	GPW-14456	Filter (Coil pri.) to junction block cable
A-5041	GPW-18846	Voltage regulator to generator ("G") cable (Bond No. 8)
A-5082	GPW-14465	Voltage regulator to junction block cable
A-1733	GPW-12250-A	Circuit breaker (Between ignition switch and fuel gauge on inst. board)
A-1734	GPW-12250-B	Circuit breaker (Between ammeter and horn)
635985	GPW-14487-A	Connector (3 wire)

If, after checking all grounds and wire connections, gauge does not indicate properly, remove wire from tank gauge unit and ground it to frame while ignition switch is on. Gauge should then read FULL. Remove wire from frame (with ignition switch on) and gauge should read EMPTY. If this is not the case, dash gauge should be replaced with a new one. If the gauge indicates as described, the trouble is probably in the tank unit, which should be replaced. Do not attempt to repair either gauge or tank unit, replacement is the only practical procedure.

Lighting System

The wiring of the lighting system is shown in Fig. 3. The lights are controlled by switches within easy reach of the driver. The lighting circuit is protected by an overload circuit breaker, which clicks off and on in the event of a short circuit in the wiring. The circuit breaker is located at the rear end of the main light switch and no replaceable fuse is required.

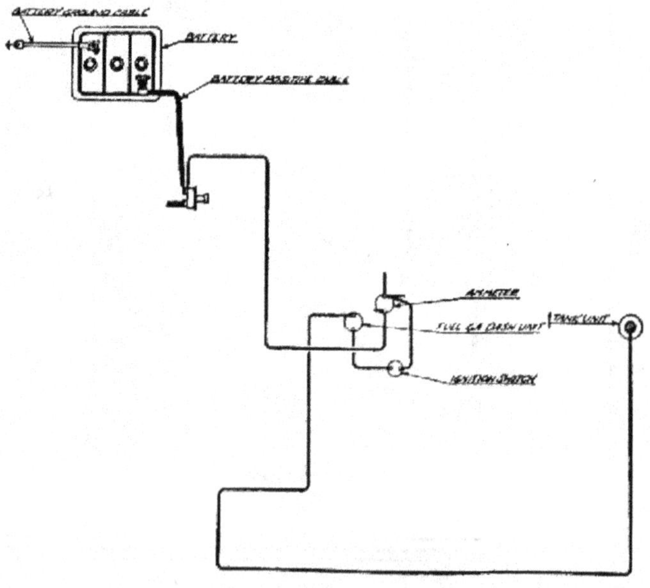

FIG. 2—FUEL GAUGE CIRCUIT

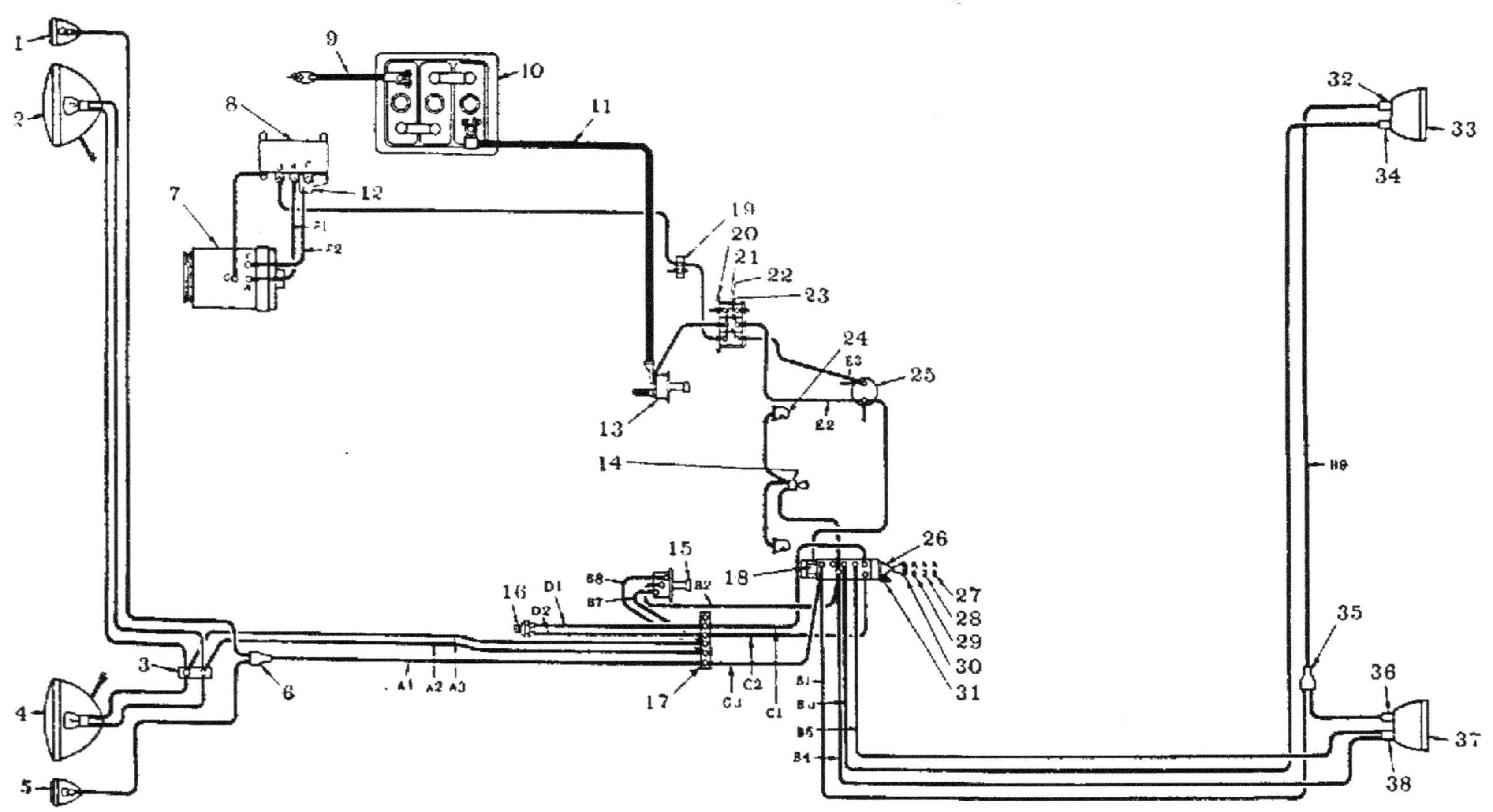

FIG. 3—LIGHTING SYSTEM

No.	Willys Part No.	Ford Part No.	Name
1	A-1437	GPW-13200	Blackout Head Lamp—Right
2	A-1305	GPW-13005	Service Head Lamp—Right
3	639599	GPW-14448-A	Junction Block—2 post
4	A-1304	GPW-13006	Service Head Lamp—Left
5	A-1436	GP-13700-B	Blackout Head Lamp—Left
6	635985	GPW-14487-A	Connector
7	A-5992		Generator
8	A-1409	GPW-10505	Voltage Regulator and Circuit Breaker
9	A-1320	GPW-14301	Battery Ground Strap
10	A-1238	11AS-10655	Battery
11	A-1452	GPW-14300	Battery Positive Cable
12	A-1287	GPW-18936-A	Radio Filter Unit (Generator Regulator)
13	A-6181		Starter Switch
14	A-1333	GPW-13740	Instrument Lamp Switch
15	638979	GPW-13532	Head Lamp Foot Switch
16	A-1271	11A-13480	Stop Lamp Switch
17	A-1490	GPW-14448-B	Junction Block—6 post
18	A-1349	GPW-12250-C	Thermal Circuit Breaker—30 Amp.
19	639599	GPW-14448-A	Junction Block—2 post

No.	Willys Part No.	Ford Part No.	Name
20	A-5980		Filter Group and Bracket Assembly
21			Filter Terminal to Battery
22			Filter Terminal to Regulator (B)
23			Filter Terminals to Ammeter
24	A-1411	GPW-13710	Instrument Lamp Socket and Cable
25	A-5231	GPW-10850	Ammeter
26	A-1332	11TS-11654	Blackout Lighting Switch
27			Service Stop Light Position
28			Service Position
29			Blackout Position
30			Off Position
31			Service Position Lock
32			Blackout Tail Light
33	A-1065	GP-13404-B2	Tail and Stop Lamp—Right
34			Blackout Stop Light
35	635985	GPW-14487-A	Connector
36			Blackout Tail Light
37	A-1064	GP-13405-B2	Tail and Stop Lamp—Left
38			Service Tail and Service Stop Light

Main Light Switch

The main light switch Fig. 4 has four positions. When the switch button is all the way in, all lights are turned off. Pulling the switch out to the first position turns on the blackout lamps, the blackout tail lamp and also connects the circuit with a blackout stop lamp on the right side which is operated through the stop light switch when the brakes are applied.

To turn on the service headlamps, it is necessary to push down on the lock-out control button and while holding it down, pull the switch button out to the next position.

During the day to cause the service stop light only, to operate upon brake application, pull the knob out to the last position. This should be done whenever the vehicle is used under ordinary driving conditions.

When installing a new light switch refer to Fig. 4 and wiring diagram, Fig. 3 which will be helpful in determining the proper wires to install on terminals as marked.

The upper and lower headlight beams are controlled by a foot switch located on the toe board at the left side.

The instrument panel lights can only be turned on when main light switch is in the service position.

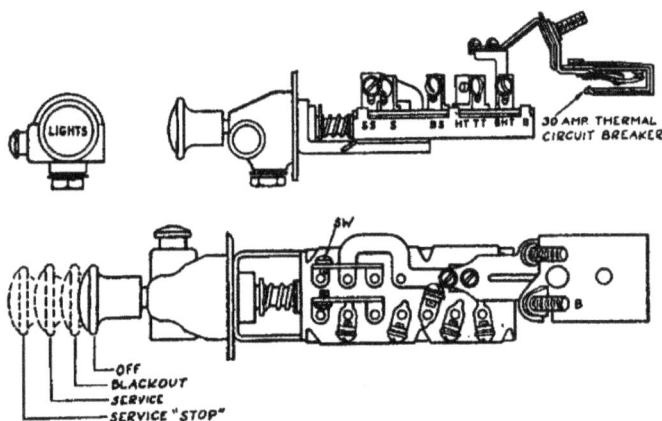

FIG. 4—MAIN LIGHT SWITCH

Stop Light Switch

Stop light switch is located in front end of brake master cylinder. The switch is a diaphragm type and closes the circuit when pressure is applied to the brakes forcing the fluid against the diaphragm which closes the circuit. When switch becomes inoperative it is necessary to install a new switch.

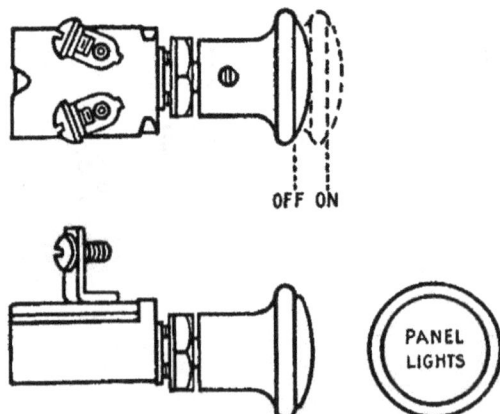

FIG. 5—PANEL LIGHT SWITCH

Panel Light Switch

The panel light switch, Fig. 5, controls the panel lights only when the main light switch is in "Service Position." Pull out on the switch knob to light the panel lights.

Head Lamps

The head lamps, Fig. No. 6, are the sealed beam type, in which the reflector, bulb and lens form a sealed unit and can only be replaced as a unit.

The lower beam filament is positioned slightly to one side of the focal point in the reflector, this results in deflecting the lower beam to the right side to illuminate the side of the road when meeting other vehicles on the highway.

To replace a burned out sealed beam unit remove door clamp screw and remove the door, No. 2, remove sealed beam assembly No. 1 and remove from connector at the rear of the unit. Install a new unit by reversing the above operations.

When a sealed beam unit has been replaced, check the aim of the head lamps.

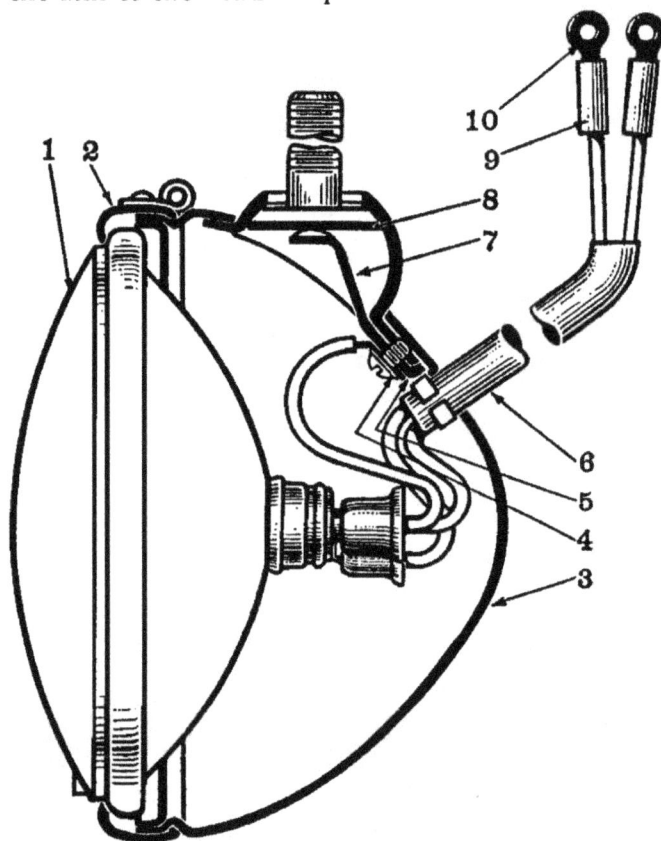

FIG. 6—HEADLAMP

No.	Willys Part No.	Ford Part No.	Name
1	A-1033	GPW-13007	Headlamp Seelite Unit Assembly
2	A-1036	GPW-13043	Headlamp Door Assembly
3	A-5586	GPW-13012	Headlamp Housing Sub—Assembly
4	A-1032	27693-S	Mounting Bolt Retainer Screw
5	52221	34803-S2	Mounting Bolt Retainer Lockwasher
6	A-1362	GPW-13076	Wire Assembly—Left (A-1363—Right)
7	A-1031	GPW-13015	Mounting Bolt Retainer
8	A-1361	GPW-13022	Mounting Bolt
9	306688	B-14455	Insulating Sleeve
10	307556	B-14463	Terminal

Head Lamp Aiming

Headlights may be aimed by use of an aiming screen or wall, Figure 7, providing a clear space of 25 feet from the front of the headlights to the screen or wall is available.

The screen should be made of a light colored material and should have a black center line for use in centering the screen with the vehicle. The screen should also have two vertical black lines, one on each side of the center line at a distance equal to lamp centers.

Place the vehicle on a level floor with the tires inflated to recommended specification. Set the vehicle 25 feet from the front of the screen or wall so that the center line of the truck is in line with the center line on the screen. To determine the center line of the vehicle, stand at the rear and sight through the windshield down across cowl and hood.

Measure from the floor to the center of the head lamp and mark a horizontal line on the screen 7" less.

Turn on the headlight upper beam, cover one lamp and check the location of the upper beam on the screen. The center of the hot spot should be centered on the intersection of the vertical and horizontal lines on the screen as shown in Fig. 7.

If aim is incorrect, loosen the nut on the mounting bolt and move the head lamp body on its ball and socket joint until the beam is aimed as described, then tighten.

Cover the head lamp just aimed and adjust the other in the same manner.

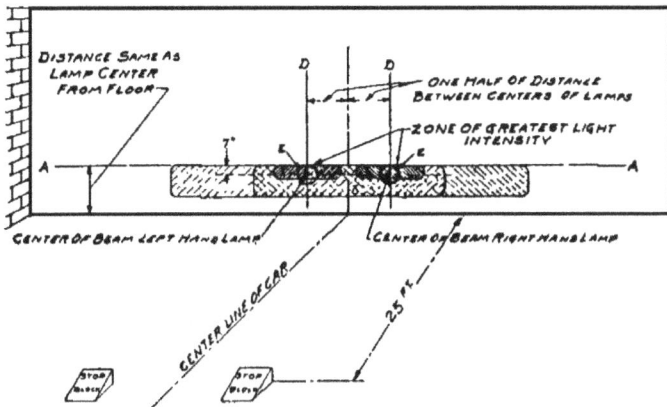

FIG. 7—HEADLIGHT AIMING CHART

Blackout Lamps

The blackout light, Fig. 8, is based on the principle of polarized light. The lens is so designed that only horizontal light beams are allowed to penetrate or pass thru the lens. This means the vertical light beams are blocked by the lens, therefore light rays cannot be seen from a point above the horizontal.

To replace lamp bulb remove door Screw No. 2 in lower side of rim—remove door No. 3 by slipping off bottom and tilt outward and up from lamp body. The door and lens are one unit. Replace Bulb (Mazda No. 63) and inspect gasket; if damaged replace and install door.

Tail and Stop Lamps

The tail and stop lamps, Fig. 9, consist of two separately sealed units placed in the Lamp Body.

The upper stop light or service unit consists of lens, gasket, reflector and (21-3 C.P. Bulb, L.H., No. 8, Fig. 9—R.H. 3 C.P., No. 4, Fig. 9) sealed as a unit. When Bulb fails entire Service unit must be replaced.

The Lower Tail lamp unit, No. 3 and 7, consists of lens, gasket, reflector and 3 C.P. Bulb sealed as a unit. When Bulb fails entire unit must be replaced.

To replace a unit remove the two screws in lamp door. Remove door then each unit can be pulled out of socket in lamp Body.

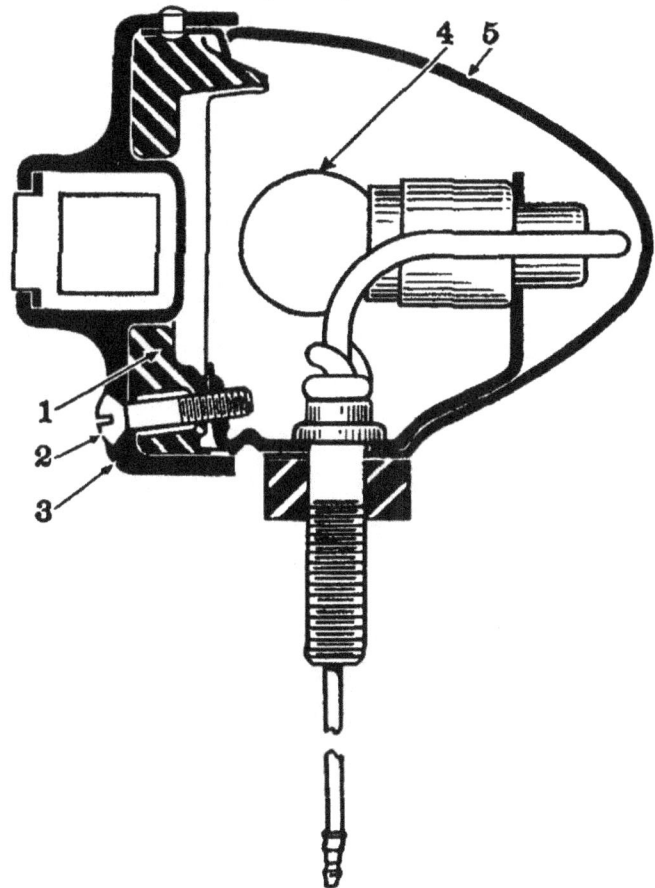

FIG. 8—BLACKOUT LAMP

No.	Willys Part No.	Ford Part No.	Name
1	A-1071	GP-13209-B2	Door Gasket
2	A-1072	28378-S2	Door Screw
3	A-1070	GP-13210-B	Door Assembly
4	51804	B-13466	Bulb
5	A-1439	GPW-13217	Housing Assembly—Left (Includes Wire Assembly) (Willys A-1440—Ford GPW-13216-B Right)

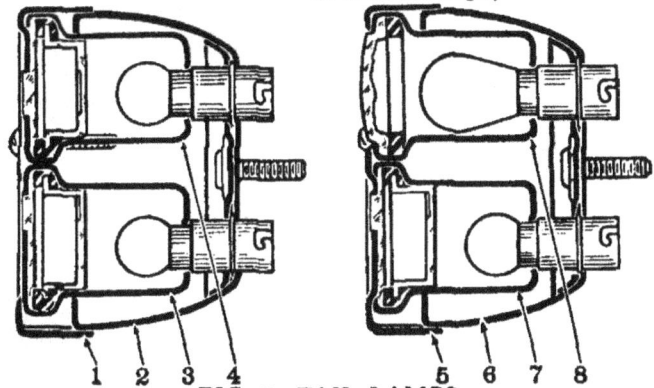

FIG. 9—TAIL LAMPS

No.	Willys Part No.	Ford Part No.	Name
1	A-1079	GP-13449-A	Door—Tail and Stop Lamp Assembly (Right)
2	A-1073	GPW-13408-B2	Housing Sub-assembly
3	A-1075	GP-13491-A2	Lower Tail Lamp Unit Assembly
4	A-1078	GP-13485-A2	Upper Stop Lamp Unit Assembly—Tail and Stop Lamp Assembly (Right)
5	A-1076	GP-13448-B2	Door—Tail and Stop Lamp Assembly (Left)
6	A-1073	GPW-13408-B	Housing Sub-assembly
7	A-1075	GP-13491-A	Lower Tail Lamp Unit Assembly
8	A-1074	GPW-13494-A2	Upper Service Assembly—Tail and Stop Lamp Assembly (Left)

IGNITION SYSTEM

The power in an internal combustion engine is derived from burning a fuel and air mixture in the engine cylinders under compression. In order to ignite these gases a spark is made to jump a small gap in the spark plug within each combustion chamber. The ignition system furnishes this spark. The spark must occur in each cylinder at exactly the proper time and the spark in the various cylinders must follow each other in sequence of firing order. To accomplish this the following parts are used:

The battery, which supplies the electrical energy;

The ignition coil, which transforms the battery current to high-tension current which can jump the spark plug gap in the cylinders under compression;

The distributor, which delivers the spark to the proper cylinders and incorporates the mechanical breaker, which opens and closes the primary circuit at the proper time;

The spark plugs, which provide the gap in the engine cylinders;

The wiring, Fig. 10, which connects the various units;

The ignition switch to control the battery current when it is desired to start or stop the engine.

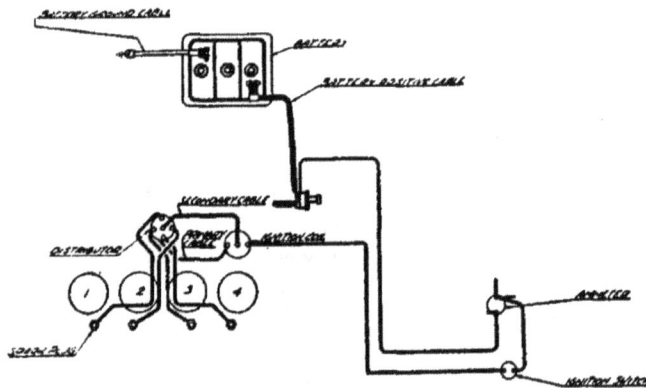

FIG. 10—IGNITION WIRING

Distributor

The distributor, Fig. 11 is mounted on the right hand side of the engine and is operated by a coupling on the oil pump shaft, driven by a spiral gear on the camshaft. The spark control is fully automatic, being operated by two counterweights pivoted on a plate which advances the timing automatically as the engine speed increases.

Distributor Overhaul

To remove distributor from engine the following procedure should be followed:

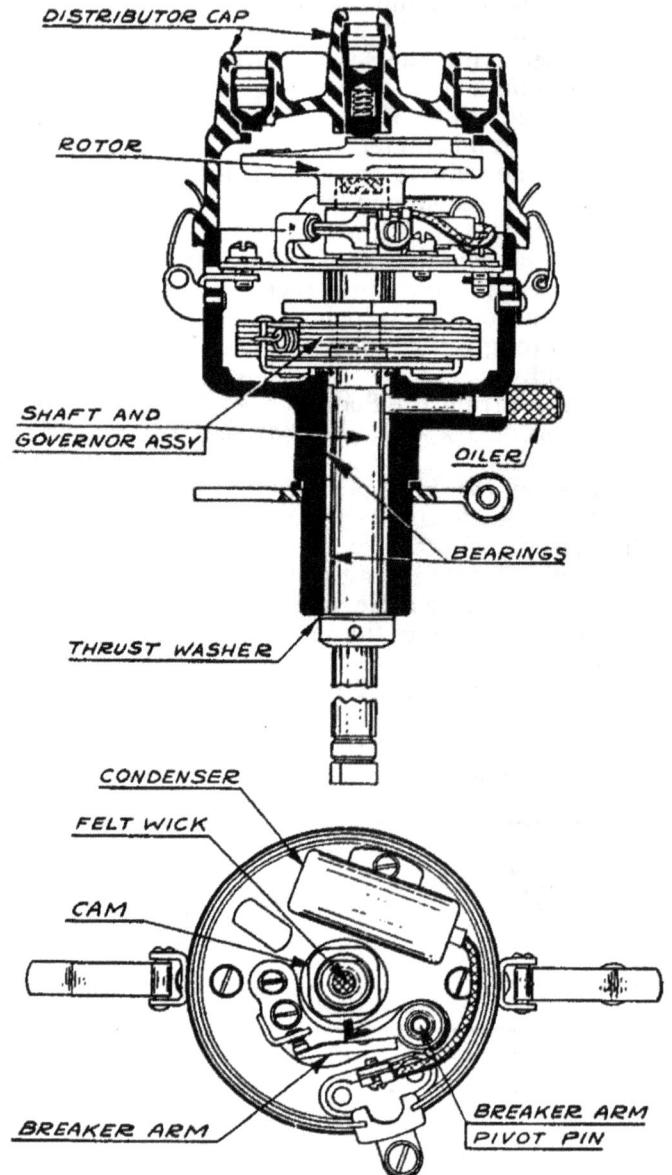

FIG. 11—DISTRIBUTOR

1. Remove high-tension wires from the distributor cap terminal towers, noting the order in which they are assembled to assure proper installation on reassembling. No. 1 spark plug terminal tower in distributor cap is the lower right hand tower at distributor cap spring clip. Starting with this tower the wires should be installed in a counter-clockwise direction 1-3-4-2.

2. Remove the primary lead from the terminal post at the side of the distributor.

3. Snap off the two distributor cap springs and lift the distributor cap off of the distributor housing.

4. Note the position of the rotor in relation to the base. This should be remembered to facilitate reinstalling and timing.

5. Remove the screw holding the distributor to the crankcase and lift the distributor from the engine.

6. Wash all parts thoroughly in a suitable cleaning fluid.

Distributor Cap

The distributor cap should be visually inspected for cracks, carbon runners, evidence of arcing, and corroded high-tension terminals. If any of these conditions exist, the cap should be replaced.

Rotor

Inspect the rotor for cracks or evidence of excessive burning at the end of the metal strip.

After a distributor rotor has had normal use, the end of the contact will become burned. If burning is found on top of the strip, it indicates the rotor is too short and needs replacing. Usually when this condition is found, the distributor cap insert will be burned on the horizontal face and the cap will also need replacing.

Distributor Points

The contacts should be clean and not burned or pitted. The contact gap should be set at .020" and should be checked with a wire gauge and re-adjusted if necessary by loosening the lock screw then turn the eccentric head screw. After adjusting, tighten the lock screw and then recheck the gap. If new contacts are installed they should be aligned so as to make contact near the center of the contact surfaces. Bend the stationary contact bracket to be sure of proper alignment and then recheck the gap.

The contact point spring pressure is very important and should be between 17 to 20 ounces. Check with spring scale hooked in the breaker arm at the contact and pull in a line perpendicular to the breaker arm. Make the reading just as the points separate. This pressure should be within the limits, too low a pressure will cause missing at high speeds, too high a pressure will cause excessive wear on the cam, block and points. Adjust the point pressure by loosening the screw holding the end of the contact arm spring and slide the end of the spring in or out as necessary. Retighten the screw and recheck the pressure.

Check the condenser, it should show a capacity of .18 to .26 microfarads. Check the condenser lead for broken wires or frayed insulation, clean and tighten the connections of the terminal posts. Be sure the condenser is firmly mounted to the distributor plate.

Governor Mechanism

The governor should be checked for free operation holding the distributor shaft and turn the cam to the left as far as it will go and release. The cam should immediately return to its original position with no drag or restrictions. Inspect the distributor shaft bearing in housing, also the shaft friction spring on end of shaft inserted into the coupling on the oil pump shaft; if damaged replace.

Setting Ignition Timing

Remove all spark plugs from engine, reinstall No. 1 spark plug finger tight. Loosen screw holding timing hole cover to flywheel housing which is located just under the starting motor on the right hand side of the engine, slide cover to side. Rotate engine crankshaft until No. 1 piston is coming up on the compression stroke, remove spark plug and rotate crankshaft slowly until the marking on flywheel "IGN" appears in the center of the timing hole in flywheel housing. Fig. 15 in Engine Section.

Place distributor rotor at No. 1 tower in distributor cap so that the points are just breaking.

Place the distributor in place on engine. When end of shaft enters driving collar on oil pump, rotate distributor shaft back and forth until driving lug on end of shaft enters the slot in coupling, then push distributor assembly down. Install holddown screw. Connect primary wire from coil to distributor. Rotate distributor body until points are just breaking, then lock in place by clamp screw. Install spark plugs and wires to distributor cap terminal towers starting with No. 1, installing in counter-clockwise direction, in the following order: 1-3-4-2. Start engine and run until it is fully warmed up, then recheck timing with Neon Timing Light. Accelerate engine and note automatic advance action.

Note: For 68 octane fuel (gasoline) set timing at top center (TC).

Generator

The generator, Fig. 12 is an air-cooled, 40 ampere, two brush type, and cannot be adjusted to increase or decrease output, as this is accomplished by use of a three-unit voltage regulator, consisting of a cutout relay, current limiting regulator and voltage regulator.

A periodic inspection should be made of the charging circuit. Under normal conditions an inspection of the generator should be made each 6,000 miles, however, the interval between these checks will vary depending upon the type of service. Dirt, dust and high speed operation are factors which contribute to increased wear of the bearings, brushes and commutator. Before assuming that any difficulty lies in the generator, a visual inspection should be made of all wiring, Fig. 13 to be sure that there are no broken wires and that all connections are clean and tight. Due attention should also be given to the Voltage Regulator, as covered under heading "Regulator" in this section. Bracket bolt torque wrench reading, 31-35 ft. lbs.

MAINTENANCE PROCEDURE

1. Commutator

If the commutator is dirty or discolored, it can be cleaned by holding a piece of No. 00 sandpaper against it while running the generator slowly. Blow the sand out of the generator after cleaning the commutator. If the commutator is rough or worn, the generator should be removed, the armature taken out, and the commutator turned down in a lathe. After turning the commutator, the mica should be undercut to a depth of $\frac{1}{32}$".

To test the armature for ground connect one prod of test set to the core or shaft, (not on bearing surfaces) and touch a commutator segment with the other. If the lamp lights, the armature winding is grounded and the armature should be replaced.

To test for short in armature coils a growler is necessary. Place the armature on the growler and hold a thin steel strip on the armature core. The armature is then rotated slowly by hand, and if a shorted coil is present, the steel strip will vibrate.

GROUND SCREW
ARMATURE TERMINAL
FIELD TERMINAL
FIELD GROUND

FAN & DRIVE PULLEY

FIELD COILS
ARMATURE

BALL
BEARING

COMMUTATOR   POLE PIECE
GROUNDED BRUSH   DRIVE END HEAD

BALL BEARING

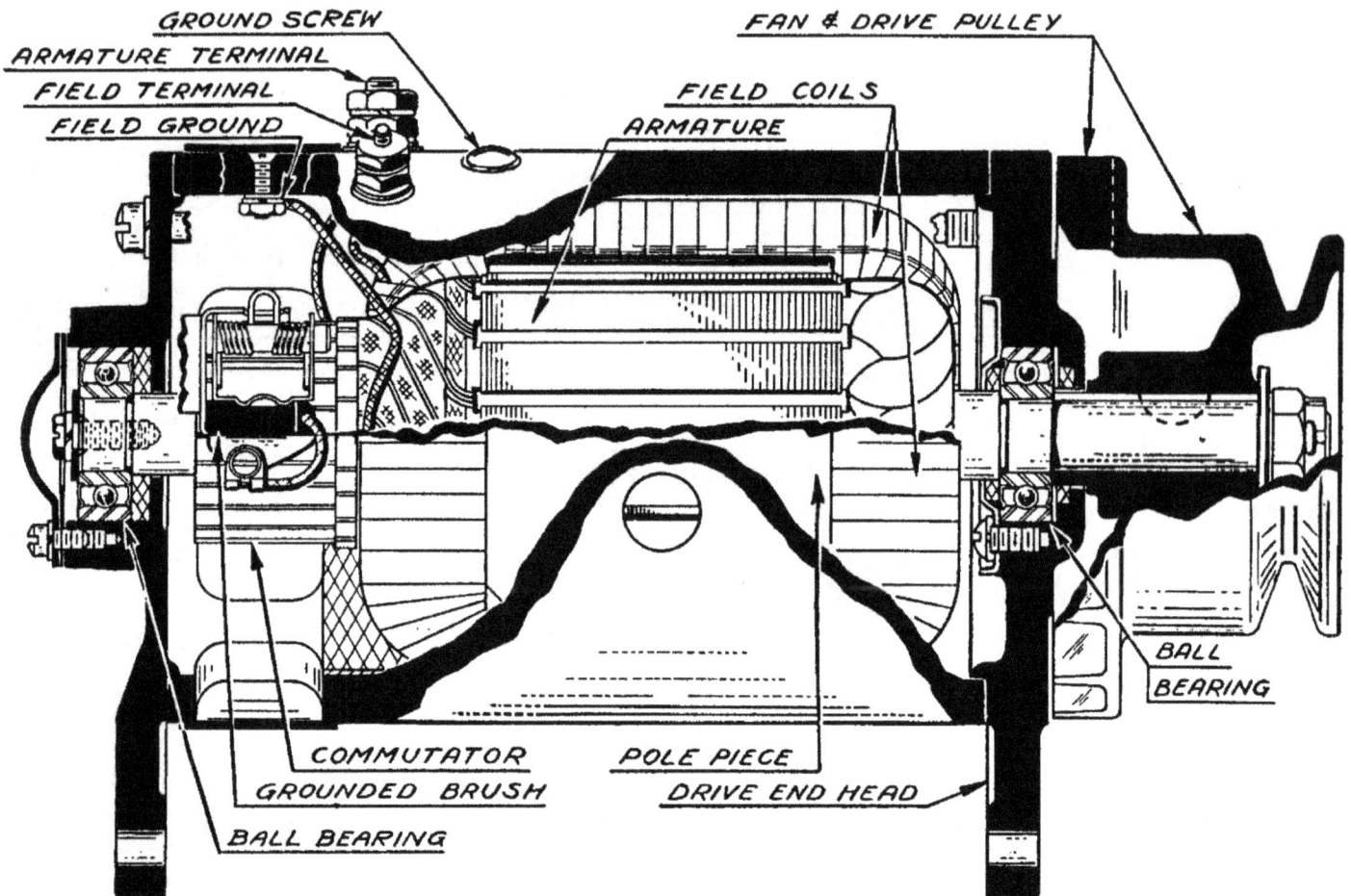

FIG. 12—GENERATOR

2. Brushes

The brushes should slide freely in their holders. If the brushes are oil soaked or if they are worn to less than one half of their original length, they should be replaced.

When replacing brushes, it is necessary to seat them so that they have 100% surface contact on the commutator. The brushes should be sanded to obtain this fit. This can be done by drawing a piece of No. 00 sandpaper around the commutator with the sanded side against the brush. After sanding the brushes, blow the sand and carbon dust out of the generator.

3. Brush Spring Tension

The brush spring tension should be checked. If the tension is excessive, the brushes and commutator will wear rapidly, while if the tension is low, arcing between the brushes and commutator will burn the commutator and reduce output. The brush spring tension is 64-68 ounces with new brushes.

4. Field Coils

Using test prods check the field coils for both open and ground. To test for open coil, connect the prods to the two leads of each coil. If the lamp fails to light, the coil is open and should be replaced.

To test for grounds, disconnect field coil ground terminal, place one prod on ground and the other on the field coil terminal. If a ground is present the lamp will light and the coil should be replaced.

5. Brush Holders

With test prods, check the insulated brush holder to be sure it is not grounded.

Touch the insulated brush holder with one prod and a convenient ground on the end plate, with the other prod. If the lamp lights, it indicates a grounded brush holder.

Inspect the brush holders for distortion and improper alignment. The brushes should swing or slide freely and should be perfectly in line with the commutator segments.

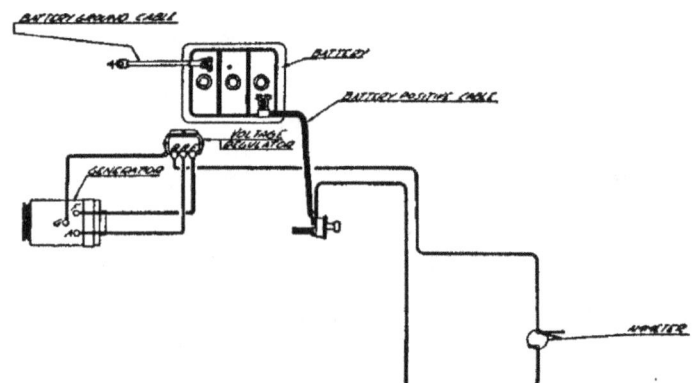

BATTERY GROUND CABLE
BATTERY
BATTERY POSITIVE CABLE
VOLTAGE REGULATOR
GENERATOR

FIG. 13—GENERATOR WIRING CIRCUIT

REGULATORS

Regulators

The generator on this vehicle is controlled by a regulator unit, Fig. 14-15 which contains a voltage regulator, current limiting regulator, and circuit breaker.

The voltage regulator controls the generator voltage and does not allow it to rise above a value determined by the voltage regulator setting. This prevents overcharging of the battery.

The current regulator controls the maximum generator output of 40 amperes and does not allow the output to exceed the value determined by the current regulator setting. This prevents damage to the generator due to an overload.

The circuit breaker automatically closes the circuit between the generator and battery when the generator voltage rises above that of the battery, and automatically opens the circuit when the generator voltage falls below that of the battery.

The terminals of the regulator unit are marked and care should be used in making connections, otherwise serious damage may result.

Quick Check

The following checks may be made to determine whether or not the units are operating normally. If not, the checks will indicate whether the generator or regulator is at fault, so that proper correction can be made:

A Fully Charged Battery and a Low Charging Rate

A fully charged battery and a low charging rate indicate normal current regulator operation. To check the current regulator, remove the battery wire from the battery terminal of the regulator. Connect the positive lead of an Ammeter to the battery terminal of the regulator and the negative lead to the battery wire with the ignition switch in the off position. Push in on the starting switch and crank the engine about 30 seconds. Then start engine and with it running at a medium speed turn on the lights and other electrical accessories and note quickly the generator output, which should be the value for which the current regulator is set.

Turn off the lights and other electrical accessories and allow the engine to continue to run. As soon as the generator has replaced in the battery the current used in cranking, the voltage regulator, if operating properly, will taper the output down to a few amperes.

A Fully Charged Battery and a High Charging Rate

Disconnect the field wire from the field terminal of the regulator. This opens the generator field circuit and the output should immediately drop off. If it does not, the generator and field wires are shorted together in the wiring harness. If the output drops off to zero with the field lead disconnected, the trouble has been isolated in the regulator. Reconnect the field lead on the field terminal of the regulator.

Remove the regulator cover and depress the voltage regulator armature manually to open the points. If the output now drops off, the voltage regulator unit has been failing to reduce the output as the battery came up to charge, a voltage regulator adjustment is indicated.

If separating the voltage regulator contact points does not cause the output to drop off, the field circuit within the regulator is shorted and the regulator should be replaced.

With a Low Battery and Low or no Charging Rate

Check the entire generator wiring circuit for loose connections, corroded battery terminals, loose or corroded ground straps. High resistance resulting from these conditions will prevent normal charge from reaching the battery and possibly cause burned out lamp bulbs when in use. If the entire charging circuit is in good condition, then either the regulator or generator is at fault.

With a jumper wire connect the field and armature terminals together, increase the generator speed and check the output. If the output increases, the regulator requires attention. If the output does not increase, a further check is necessary.

If the generator output remains at a few amperes with the field and armature terminals connected together, the generator is at fault and should be checked.

If the generator does not show any output at all, either with or without the field and armature terminals together, flash the armature terminal on the generator to ground with a screw driver or a pair of pliers, with the generator operating at medium speeds. If a spark does not occur the trouble has been isolated in the generator and it should be removed and repaired. If a spark does occur, likely the generator can build up, but the circuit breaker is not operating to let the current flow to the battery, due to burnt points, points not closing, open voltage windings, or too high a voltage setting of the cutout.

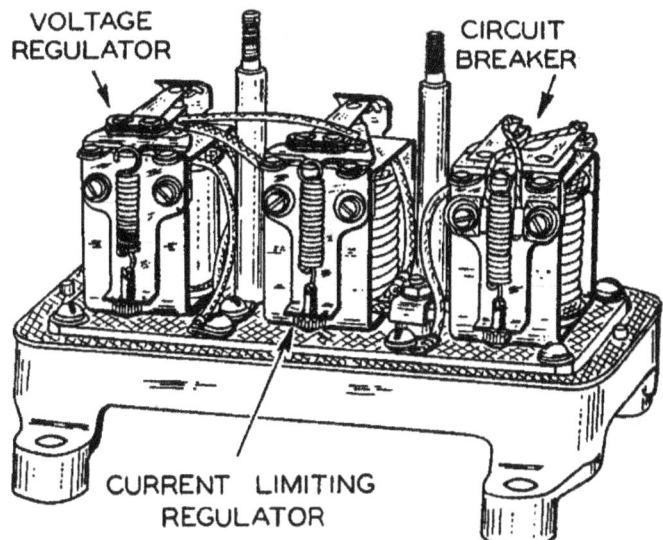

FIG. 14—VOLTAGE REGULATOR

Adjustment

To accurately adjust the regulator, it is essential to use a precision ammeter, voltmeter, thermometer and a fully charged battery. If battery in vehicle is not fully charged, it should be substituted with a fully charged battery in order to obtain a satisfactory setting of the unit.

The engine should be started and allowed to run for a period of fifteen minutes at approximately 25 to 30 miles per hour with the hood up before taking any meter readings.

The thermometer should be placed so that the bulb is approximately two inches from the side of the regulator so that the temperature can be taken while checking units. Set engine speed so that the generator charges 20 amperes, the voltmeter should show a reading according to the following specifications:

Temperature Fahr.	Volts
50°	7.41
60°	7.38
70°	7.35
80°	7.32
90°	7.29
100°	7.26
110°	7.23
120°	7.20

Tolerance—plus or minus .15 volts.

Circuit Breaker

Disconnect the battery wire from the battery terminal of the regulator. Connect the positive lead of the ammeter to the battery terminal of the regulator and the negative lead to the battery wire. Connect the positive lead of the voltmeter to the armature terminal of the regulator and the negative lead to the ground or regulator base. Gradually increase the engine speed, noting the voltage at which the points close. This should be 6.4 to 6.6 volts. Slowly decrease the engine speed, noting the discharge of current necessary to open the cutout points. This should be 0.5 to 6.0 amperes.

Check the armature air gap with the points open. Use a flat gauge .0595" to .0625". Insert between magnet core and the armature on contact side of brass pin in core. To adjust, bend armature stop.

Check the gap of the contact points when open. This should be .015" minimum, but will possibly be more than this in actual adjustment. Adjust by bending the supporting arms of the stationary points, be sure that the points are perfectly aligned.

The closing voltage of the circuit breaker may be adjusted by adjusting the screw holding the lower end of the spring. The point opening amperage can be adjusted by raising or lowering the stationary points by bending the supporting arms of the points. Be sure that there is a minimum point gap of .015".

Voltage Regulator

Connect the positive lead of voltmeter to the battery terminal and the negative lead to a ground on regulator base.

Run the engine at a speed equivalent to approximately 30 miles per hour and check the voltmeter reading, which should be in accordance with the specifications of temperature and volts (under

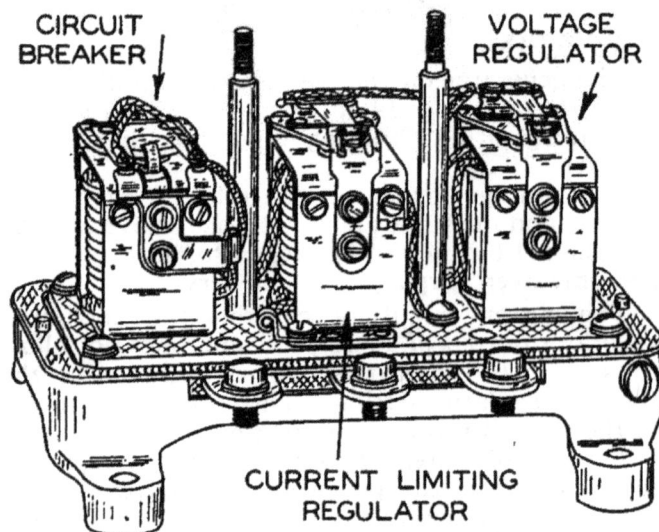

CIRCUIT BREAKER   VOLTAGE REGULATOR

CURRENT LIMITING REGULATOR

FIG. 15—VOLTAGE REGULATOR

heading "Adjustments") with a generator charging rate of 20 amperes.

Check armature air gap using .040" to .042" pin gauge. Test with pin gauge between magnet core and armature. This measurement should be taken on the contact side and next to the brass armature stop pin. To test connect a 3 candlepower test light in series with the armature and field terminals and a battery. With the low limit pin gauge in place depress the armature and the light should go out. With the high limit pin gauge in place depress the armature and the light should stay lit. To adjust slightly loosen the screw holding the upper point bracket, raise or lower bracket until correct gap is obtained. Keep points in perfect alignment when adjusting.

Check and see that the spring upon which the movable contact is mounted is straight and that it is approximately parallel with the armature.

The gap between the contact spring (upper) and armature stop is .010" to .016" with armature depressed.

Check the point gap with armature against stop pin. Hold the armature down with two fingers being careful not to apply pressure to the spring supporting the upper point. A .010" (minimum) feeler gauge should be used between points. Too much variation indicates wrong length of the brass armature stop pin and a new unit will be required.

To adjust the regulator increase or decrease the armature spring tension by adjusting the screw which holds the lower end of the spring.

Current Limiting Regulator

To adjust current limiting regulator, remove the battery wire from the battery terminal of the regulator and connect the positive lead of ammeter to the battery terminal of the regulator and the negative lead to the battery wire. Turn on the lights and other electrical accessories, then increase the engine speed until output remains constant. The ammeter reading with the unit at operating temperatures should be 40 amperes.

To check the armature air gap and point gap—refer to instructions under heading "Voltage Regulator." The armature air gap is .047" to .049". The contact point gap is .010" minimum. The

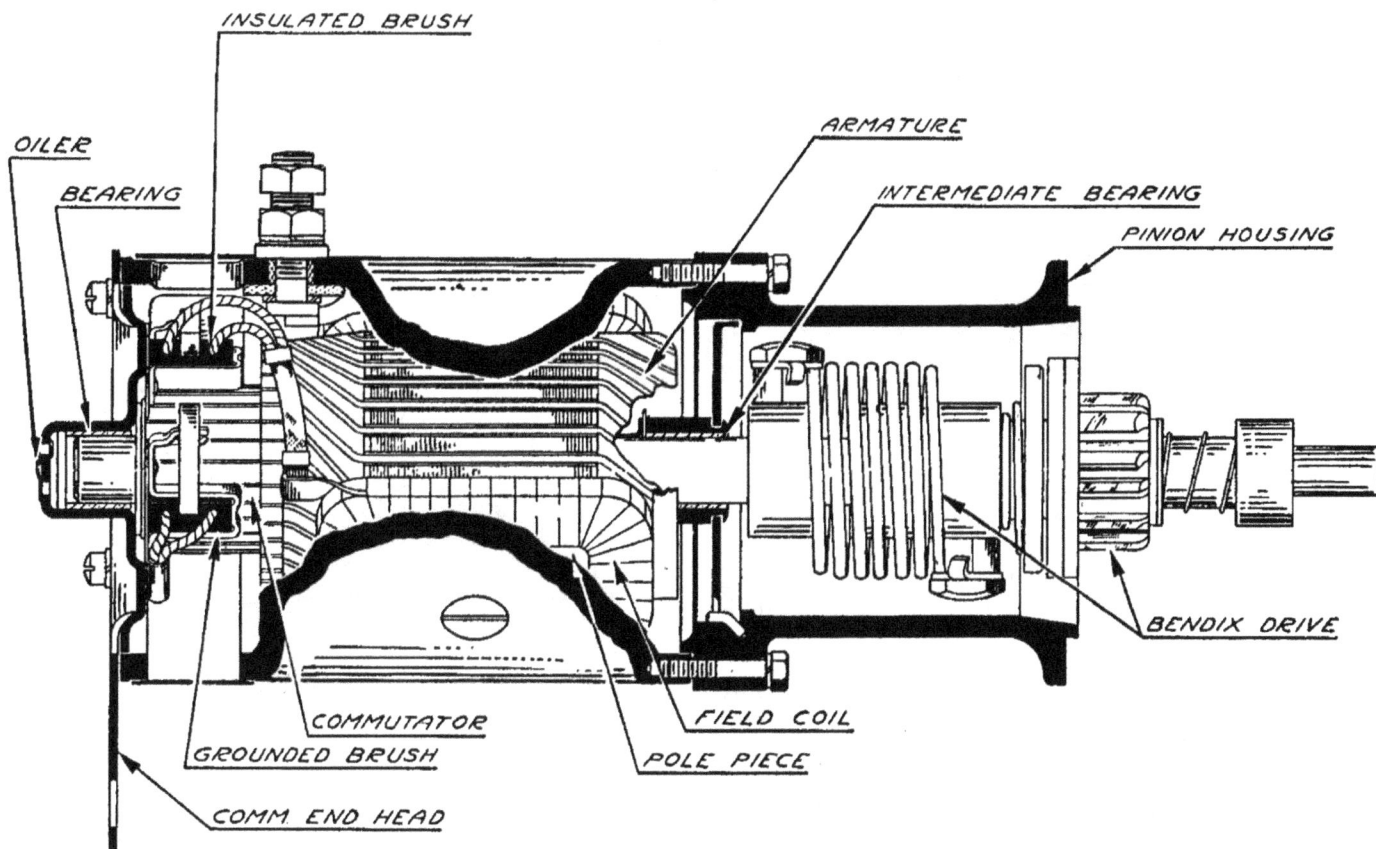

FIG. 16—STARTING MOTOR

gap between the contact spring and armature stop is .010″ to .016″ with armature depressed.

To adjust current setting vary the armature spring tension by adjusting the screw which holds the lower end of the armature spring.

STARTING MOTOR

The starting motor Fig. 16 is similar in construction and in appearance to the generator, but the design of the parts are different. Both motor and generator require a frame, field coils, armature, brushes.

A starting motor of this type requires very little attention except regular lubrication and periodic inspection of the brushes and commutator.

A visual inspection should be made of all wires and see that all connections in the circuit are clean and tight, Fig. 17. Mounting screws, torque wrench reading, 31-35 ft. lbs.

1. **Commutator**
   Check the commutator for wear or discoloration. If found to be dirty or discolored, it can be cleaned with No. 00 sandpaper. Blow the sand out of the motor after cleaning the commutator. If the commutator is rough or worn, the armature should be removed and the commutator turned down in a lathe.

2. **Brushes**
   The brushes should slide or swing freely in their holders and make full contact on the commutator. Worn brushes should be replaced.

3. **Brush Spring Tension**
   This tension should be 42 to 53 ounces with new brushes. Measure the tension with a spring scale hooked under the brush spring at end, and pull on a line parallel to the face of the

brush taking the reading just as the spring leaves the brush.

4. **Armature ·**
   The armature should be visually inspected for mechanical defects before being checked for shorted or grounded coils.

   For testing armature circuits, it is advisable to use a set of test prods.

   To test the armature for grounds, touch one prod to a commutator segment and touch the core or shaft with the other prod. Do not touch the points or prods to the bearing or brush surface, as the arc formed will burn the smooth finish. If the lamp lights, the coil connection to the commutator segment is grounded.

   To test for shorted armature coils, a growler is necessary. Place the armature on growler with a steel strip held on the armature core, rotate the armature slowly by hand. If a shorted coil is present, the steel strip will become magnetized and vibrate.

   If an armature is shorted or grounded, it will be necessary to install a new armature.

5. **Field Coils**
   Using same test prods, check the field coils for both open circuit and ground. To test for grounds, place one prod on the motor frame or pole piece and touch the other to the field coil terminal. If a ground is present, the lamp will light.

   To test for open circuit, place the prods on the field coil terminal and across each coil separately. If the lamp does not light, the coil circuit is open.

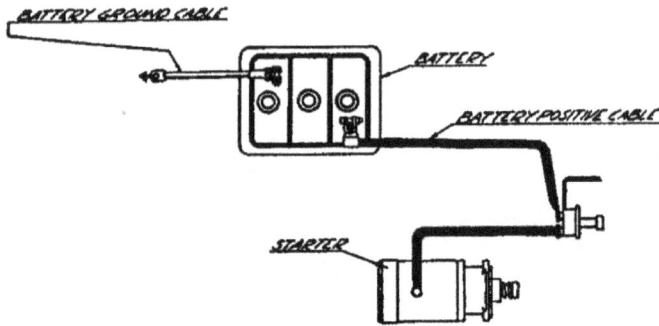

FIG. 17—STARTING MOTOR WIRING CIRCUIT

6. Brush Holder

Using test prods, touch the insulated brush holder with one prod and a convenient ground on the plate with the other. If the lamp lights, it indicates a grounded brush holder and a new brush holder will have to be installed.

Bendix Assembly

The Bendix Drive Fig. 18 is designed so that when the starting motor is energized, centrifugal force sends the counter-weighted drive pinion gear into engagement with the teeth on the flywheel. When the engine starts and the speed of the engine exceeds the comparable speed of the starting motor, the Bendix Drive pinion is forced out of engagement with the flywheel.

There are two types of Bendix Drives and springs, right hand and left hand. The type used on this starting motor is of the right hand type.

To determine right or left hand Bendix Drive, turn drive pinion so that the threads on shaft will show, note the spiral of the thread; right hand spiral, right hand drive; left hand spiral, left hand drive.

To determine a right or left hand spring, note the spiral of the coil; if to the right, it is a right hand spring; if to the left, it is a left hand spring.

If upon inspection of the Bendix Drive, the spring shows signs of being distorted, a new spring should be installed.

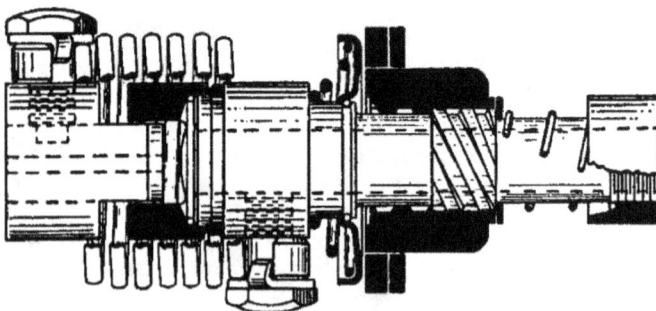

FIG. 18—BENDIX ASSEMBLY

Starting Switch

The starting switch is mounted on the toe board to the right of the accelerator pedal; pressing the starter switch closes the starter circuit and operates the starter. If the starting motor does not rotate, then the difficulty is probably a loose wire, poor ground, low battery or poor brush contact.

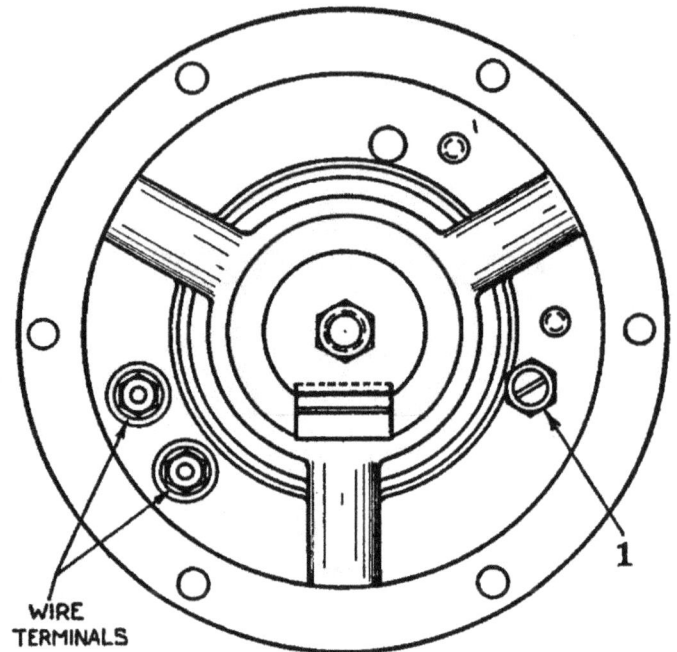

FIG. 19—HORN (Rear)

Horn and Horn Wire

The horn, Fig. 19 is the micro-vibrating type mounted on the dash under the hood. No. 1 indicates the horn adjusting screw. To adjust tone of horn, loosen the lock nut and turn the screw until the proper tone is obtained. It is best to have the engine running so the generator is charging when making this adjustment because the generator delivers 8 volts as compared to the battery 6 volts. This affects the horn tone.

The horn wire through the steering post connects to an insulator sleeve with brush contact where horn wire attaches to jacket tubing.

Whenever it is necessary to replace the horn wire, it will be necessary to remove the steering post jacket tubing. The wire may be removed by unsoldering it from the contact sleeve on the steering tube. When replacing the wire be sure to use a non-corrosive soldering flux when soldering the wire to the contact sleeve.

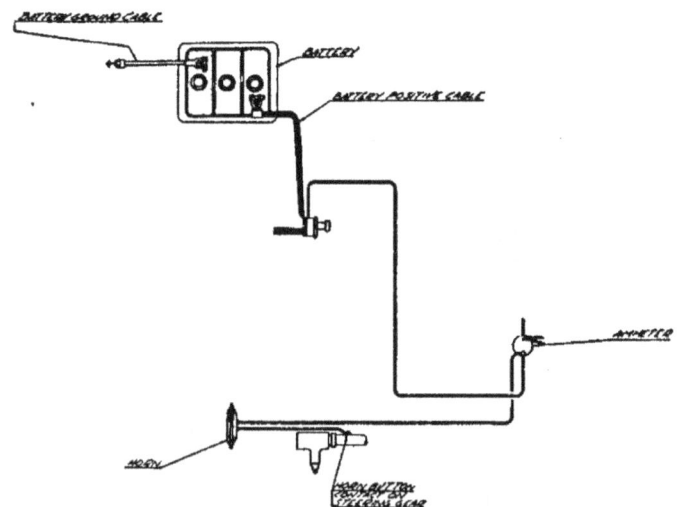

FIG. 20—HORN WIRING CIRCUIT

ELECTRICAL TROUBLES AND REMEDIES

SYMPTOMS	PROBABLE REMEDY

Battery Discharged:

Short in Battery Cell	Replace Battery
Short in Wiring	Check Wiring Circuit
Generator Not Charging	Inspect Generator and Fan Belt
Loose or Dirty Terminals	Clean and Tighten
Excessive Use of Starter	Tune Engine
Excessive Use of Lights	Check Battery

Generator:

Low Charging Rate—

Dirty Commutator	Clean Commutator
Poor Brush Contact	Install new Brushes
Voltage Regulator Improperly Adjusted	Adjust
High Resistance in Charging Circuit	Clean and Tighten Terminals
Ground Strap Engine to Frame Broken	Replace
Loose or Dirty Terminals	Clean and Tighten

Too High Charging Rate

Current Regulator Improperly Adjusted	Adjust
Short in Armature	Replace

Starting Motor:

Slow Starter Speed

Discharged Battery or Shorted Cell	Recharge
Ground Strap Engine to Frame	Clean Terminals and Tighten
Loose or Dirty Terminals	Clean and Tighten
Dirty Commutator	Clean With No. 00 Sand Paper
Poor Brush Contact	Install New Brushes
Worn Bearings	Replace
Burnt Starter Switch Contacts	Replace Switch

Distributor:

Hard Starting:

Distributor Points Burnt or Pitted	Clean Points or Replace
Breaker Arm Stuck on Pivot Pin	Clean and Lubricate
Breaker Arm Spring Weak	Replace
Points Improperly Adjusted	Adjust .020"
Spark Plug Points Improperly Set	Adjust .030"
Spark Plug Wire Terminals in Distributor Cap Corroded	Clean
Loose Terminals	Check Circuit
Loose or Dirty Terminals Ground Strap Engine to Frame	Clean and Tighten
Condenser Defective	Replace Condenser
Improper Ignition Timing	Set Timing

Lights:

Burn Dim

Loose or Dirty Terminals	Clean and Tighten
Leak in Wires	Check Entire Circuit for Broken Insulation
Poor Switch Contact	Install New Switch
Poor Ground Connection	Clean and Tighten
Aim Headlamp Beams	Use Chart

Horn Fails to Blow

1.	Broken or loose electrical connection	1. Check wiring and connections at horn button and battery, making sure all are clean and tight.
2.	Battery low or dead	2. Check battery with hydrometer, should read at least 1200.
3.	Contact points in horn not adjusted	3. Loosen locknut and turn adjusting screw to right or left until a clear steady tone is obtained, then tighten locknut, holding screw in proper position.
4.	Contact points burnt or broken off	4. Replace parts necessary and adjust horn.

Horn Blows Unsatisfactory Tone

1. Poor electrical connection.....................
1. Check wiring and connections at horn, horn button and battery.

2. Battery low...............................
2. Check with hydrometer, should read 1200.

3. Loose cover and bracket screws................
3. Draw cover screws and center nut tight, tighten bracket screws solidly both at horn and dash.

4. Voltage at horn too high or too low ....
4. Check with voltmeter, should measure 5.5-6.5 volts at horn with horn sounding and engine running so generator is charging battery.

5. Contact points are not properly adjusted........
5. Loosen locknut and turn contact adjusting screw to right or left until a clear steady tone is obtained, then tighten locknut.

Excessive Radio interference

1. Due to Ignition ....
1. Check distributor, spark plugs and suppressors.
2. Tighten braided bonding straps.
3. Tighten radiator and fender supporting bolts.

2. Due to Generator ....
1. Tighten regulator to generator bond.
2. Defective commutator, brushes or holders.
3. Discharge battery causing high discharging rate.

3. Due to Erratic Noises ....
1. Failure of high tension insulation.
2. Loose wiring connections or corroded distributor cap towers.
3. Defective switches or gauges.

ELECTRICAL SYSTEM SPECIFICATIONS

Battery:
 Make.................Auto-Lite or Willard
 Model..............TS-2-15 or SW-2-119
 Plates per Cell........................15
 Capacity....................116 Amp. Hr.
 Volts.................................6
 Length....................Approx. 10"
 Width......................Approx. 7"
 Height....................Approx. 8⅝₁₆"
 Specific Gravity:
  Fully Charged...............1225-1300
  Recharge at....................1175
 Ground Terminal................Negative
 Location..........Under Hood Right Side

Starting Motor:
 Make.........................Auto-Lite
 Model....................MZ-4113
 Drive..........Right hand outboard Bendix
 No Load Draw...............70 amps. max.;
    5.5 volts—4300 R.P.M. Min.
 Stall torque....420 amps.; 3.0 volts—7.8 ft. lbs.
 Volts.................................6
 Armature End Play.............¹⁄₁₆" Max.
 Brushes..............................4
 Brush Spring Tension............42-53 Oz.
 Normal Engine Cranking Speed....185 R.P.M.
 Bearings................3 absorbent bronze

STARTER SWITCH:
 Make.....Auto-Lite   Model.....SW-4001

Generator:
 Make.........................Auto-Lite
 Model....................GEG-5002D
 Volts.............................6-8

Ground Polarity...................Negative
Controlled Output................40 Amps.
Rotation (Drive End)...........Clockwise
Control. Vibrating type current-voltage regulator
Air Cooled.........................Yes
Armature End Play..............010" Max.
Brushes...............................2
Brush Spring Tension............64-68 Oz.
Bearings..........................Ball
Field Coil Draw..1.60 to 1.78 Amps.—6.00 V.
Motorizing Draw..4.7 to 5.2 Amps.—6.0 volts
(Have field and armature terminals connected).
Output.8.0 amps.; 7.6 volts; 955 Max. R.P.M.
    40.0 amps.; 7.6 volts; 1460 Max. R.P.M.
    40.0 amps.; 8.0 volts; 1465 Max. R.P.M.

Current—Voltage Regulator
 Make.....Auto-Lite   Model..VRY-4203A
 Volts.................................6
 Amperes..............................40
 Ground Polarity...................Negative

Voltage Regulator:
 Voltage Setting Open Circuit.........7.20-7.41
 Air Gap.......................040"-.042"
 Point Gap....................010"-.012"

Circuit Breaker:
 Points Close (Hot)............6.4-6.6 Volts
 Points Open—Reverse Current...0.5-6.0 Amps.
 Air Gap....................0595"-.0625"
 Points Gap........................015"

Current Limiting Regulator:
 Air Gap.......................047"-.049"
 Point Gap....................030"-.033"

ELECTRICAL SYSTEM SPECIFICATIONS
(Continued)

Distributor:

Make.....................Auto-Lite
Model.....................IGC-4705
Type Advance.................Centrifugal
Firing Order.................1-3-4-2
Breaker Point Gap...................020"
Breaker Arm Spring Tension........17-20 Oz.
Cam Angle (Time points are closed).......47°
Max. Automatic Advance 1500 R.P.M. (dist.).11°
Condenser Capacity..............18-.26 Mfd.
Timing—72 octane fuel (gasoline)
    5° BTC Flywheel (.0103" Piston travel)
Timing—68 octane fuel (gasoline)
    TC Flywheel (Zero Piston travel)
Timing Mark.....................Flywheel
Location..Right Side Bell Housing under Starter
Ignition Switch (Lock)......Douglas No. 5941

Coil:

Make.....................Auto-Lite
Model.....................IG-4070-L
Draw Engine Stopped.....5 Amps. @ 6.4 Volts
Draw Engine Idling..............2.5 Amps.

Gauges:

Fuel Gauge....................Auto-Lite
Oil Pressure...................Auto-Lite
Temperature....................Auto-Lite
Ammeter.....................Auto-Lite

Spark Plugs:

Make.....................Champion QM-2
Size.....................14MM
Gap.....................030"

Radio Filters:

Generator Filter Unit......Tobe Deutschmann
Regulator Filter..........Tobe Deutschmann
Filter Group............Tobe Deutschmann

Lamps:

Light Switch Make.................Douglas
Foot Beam Switch Make.......Clum No 9654
Head Lamps.....Corcoran-Brown Sealed Beam
Black Out Lamps..........Corcoran-Brown
Tail and Stop Lamp.........Corcoran-Brown
Head Lamp Bulbs (Seelite unit).6-8V-45 C.P. DC.
    Mazda No. 2400
Blackout Bulbs...6-8V-3C.P. SC Mazda No. 63
Tail and Stop Lamp Bulbs.6-8V 3-21CP Mazda
    No. 1154
Instrument Lamp Bulbs...6-8V 3CP SC Mazda
    No. 63

Horn:

Type.....................Micro-Vibrator
Make.....................Sparks-Withington
Model.....................B-9427

BONDED POINTS

Bond No.	Name	Bond No.	Name
1.	Hood to Dash—Right Hand	13.	Radiator Right Hand to Frame
2.	Hood to Dash—Left Hand	14.	Radiator Left Hand to Frame
3.	Cylinder Head Stud to Dash	15.	Rear Engine Support to Frame Cross Member Stud
4.	Brake Cable, Speedometer Cable, Heat Indicator Cable to Dash	16.	Transfer Case to Body Floor Stud
5.	Gas Line to Dash	17.	Right Hand Body Bracket Ground to Frame
6.	Choke Control, Throttle Control and Oil Gauge Line to Dash Stud	18.	Left Hand Body Bracket Ground to Frame
7.	Generator Mounting Bolt to Starting Motor Bracket	19.	Right Hand Fender Ground to Frame
8.	Generator Voltage Regulator Filter & Ground	20.	Left Hand Fender Ground to Frame
9.	Coil to Cylinder Block	21.	Left Hand Hood Ground to Grill
10.	Right Hand Front Motor Bracket to Frame	22.	Right Hand Hood Ground to Grill
11.	Left Hand Front Motor Bracket to Frame	23.	Headlamp Wiring Harness to Left Fender
12.	Exhaust Pipe to Frame	24.	Cylinder Head Stud—Front
		25.	Left Hand Fender to Cowl—Lower
		26.	Right Hand Fender to Cowl—Lower

IMPORTANT: Where parts are grounded, particular attention must be given to any special position of lockwashers on bolts and screws. Tinned spots should be clean but not painted, to assure satisfactory bond.

TRANSMISSION

The transmission, Fig. 1 is of the three speed, synchromesh type with synchronized 2nd and high speed gears. See shifting diagram, Fig. 4 in Driver's Instructions Section.

The transmission is bolted to the rear face of the flywheel bell housing with four cap screws and is supported on a rubber insulator at the center frame cross member which is the rear engine support or mounting.

Removal of Transmission and Transfer Case from Engine

1. Remove front and rear propeller shafts at universal joint in accordance with instructions under the "Propeller Shafts and Universal Joints".

2. Disconnect speedometer cable at transfer case.

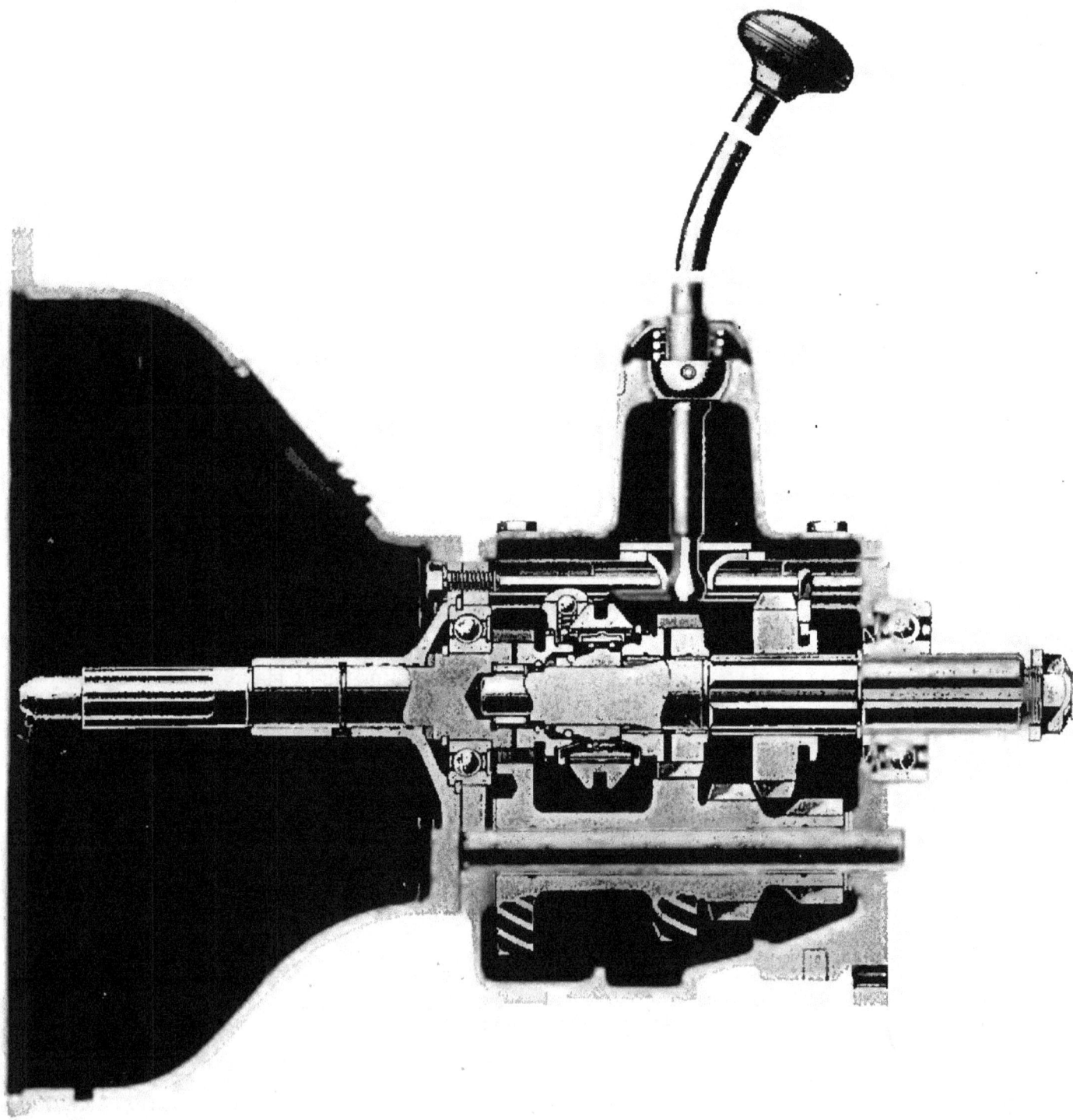

FIG. 1—TRANSMISSION

3. Disconnect brake and engine snubbing cables.

4. Remove nuts holding rear mounting to frame cross member.

5. Remove transfer case snubbing rubber bolt nut at cross member.

6. Remove transmission shift lever by unscrewing retainer collar at top of shift housing.

7. Disconnect the clutch release cable at bell crank and remove; also, remove clutch release lever No. 10, Fig. 2, "Clutch Section" through the inspection hole in the flywheel bell housing.

8. Place jacks under engine and transmission.

9. Remove floor board inspection plate, drain radiator and remove upper hose.

10. Remove transfer shift lever pivot pin screw and lubricator.

11. Remove shift lever pin and remove levers.

12. Remove bolts holding center cross member to frame side rail and remove cross member.

13. Remove bolts holding transmission to flywheel bell housing.

14. Force transmission to right and disconnect clutch control lever tube ball joint.

15. Lower jacks under engine and transmission; slide transmission assembly towards rear of vehicle until clutch shaft clears bell housing.

16. Lower jack under transmission and remove assembly from under chassis.

FIG. 2—REMOVING BEARING SNAP RING

Disassembly of Transmission

Drain lubricants from both the transmission and transfer case through drain plug holes in bottom of each case. It is advisable to clean the outside of the cases thoroughly with water or other suitable cleaning fluid before attempting to disassemble the units.

To disassemble the unit the following procedure is recommended:

1. Remove cap screws and lock washers holding rear cover. No. 37, Fig. 3 in Transfer Case Section.

2. Remove cotter pin, nut and washer permitting removal of main shaft gear No. 57.

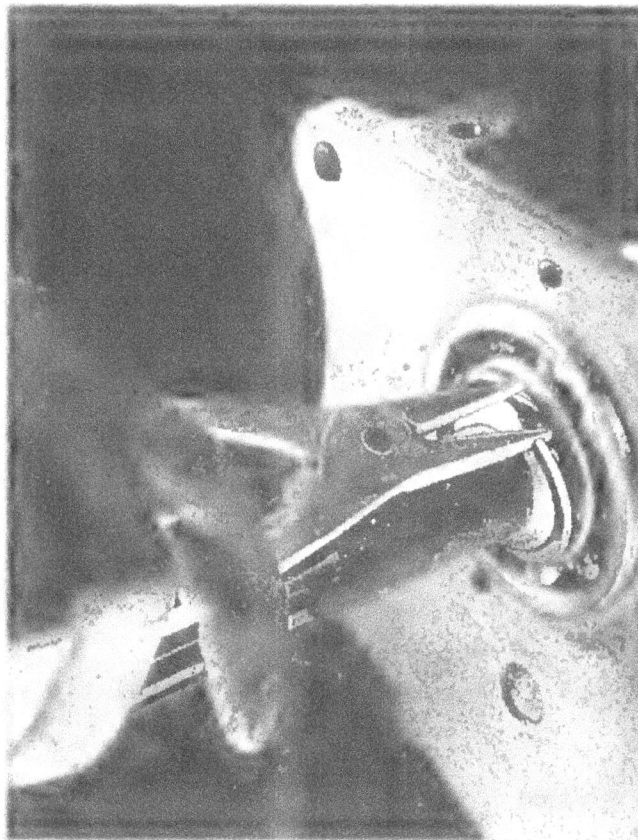

FIG. 3—REMOVING SNAP RING

3. Remove the four cap screws holding control housing to top of transmission and remove housing.

4. Remove shifter plate spring and take off shifter plate, No. 11, Fig. 4.

5. Loop a piece of wire around main shaft, just rear of main shaft second speed gear, twist wire and attach one end to the right hand front cover screw and the other end to the left hand cover screw, drawing the wire tightly to prevent the main shaft from sliding out of the case when transfer case is removed.

6. Remove the five cap screws holding the transfer case to the rear face of the transmission.

7. Support transfer case and with a rawhide mallet or brass rod and hammer, tap lightly on end of shaft and at the same time draw the transfer case away from the transmission. Be careful not to lose transmission gear shift interlock plunger. The transmission main shaft rear bearing No. 34, Fig. 4 should slide out of transfer case and remain in transmission.

8. Remove three screws holding main drive gear bearing retainer, No. 1 and remove retainer and gasket.

9. Remove shift fork guide pin, No. 20 through front of transmission.

10. Remove shift fork set screws with special wrench furnished in tool kit and remove shift shafts and forks. Be careful not to lose poppet springs and balls.

11. Remove lock plate at rear of transmission holding countershaft and reverse idler gear shaft.

12. With a drift, drive out the countershaft.

13. Remove main drive gear bearing, shaft and synchronizer blocking ring.

14. Remove snap rings from main drive gear shaft and bearing, Fig. 2 and 3 and remove bearing from shaft.

15. Remove main shaft assembly.

16. Remove countershaft gear set and three thrust washers, two bushings and a spacer.

17. Remove reverse idler gear shaft and gear.

To remove the gears on main shaft, first remove snap ring No. 27, Fig. 4, on end of shaft holding transmission high and intermediate clutch hub, No. 28. After the removal of the snap ring, the gears will slide off the shaft without difficulty. To disassemble synchronizer unit, push apart.

Wash all parts in suitable cleaning fluid and inspect for wear and damaged parts, replacing any parts which show excessive wear or damage.

Assembly of Transmission

The assembly of the parts in the transmission should be performed in the reverse manner in which it was dismantled making reference to exploded views of parts as shown in Fig. 4, for sequence of assembly.

When assembling synchronizer unit assembly place the right end of a synchronizer spring No. 14 in one shifting plate. Turn the unit around and make exactly the same installation with the other spring in the same shifting plate. This will actually place the spring action opposed to each other.

The bushings in countershaft gear set are of the floating type, being free to turn within the gear as well as on the shaft. When making assembly of countershaft to transmission case, dip these bushings in lubricant of S.A.E. 90 grade and be sure the spacer is installed between the two bushings. The steel thrust washer, No. 43 at the rear of countershaft gear is pinned in the case and the bronze washer, No. 42 is installed between steel washer and gear. Only one bronze washer, No. 37 is used at the front. The main shaft ball bearing, No. 34 is assembled to shaft so that the sealed side is in transmission, open side to transfer case.

TRANSMISSION TROUBLE AND REMEDIES

SYMPTOMS	PROBABLE REMEDY
Slips Out of High Gear	
Transmission misaligned with Bell Housing.....	{Align Transmission Case to Bell Housing and Bell Housing to Engine
End play in Main Drive Gear................	Tighten Front Retainer
Damaged Pilot Bearing or Front Bearing......	Replace
Bent Shifting Fork........................	Replace
Slips Out of Second	
Bent Shifting Fork........................	Replace
Worn Gear................................	Replace
Weak Poppet Spring.......................	Replace
Noise in Low Gear	
Rear Ball Bearing Broken..................	Replace
Gear Teeth Pitted or Worn.................	Replace gears
Shifting Fork Bent........................	Replace
Lack of Lubrication.......................	Drain and Refill
Grease Leak into Bell Housing	
Gasket Broken Front Bearing Retainer........	Replace
Transmission Case Overfilled with Lubricant...	Drain off to proper level

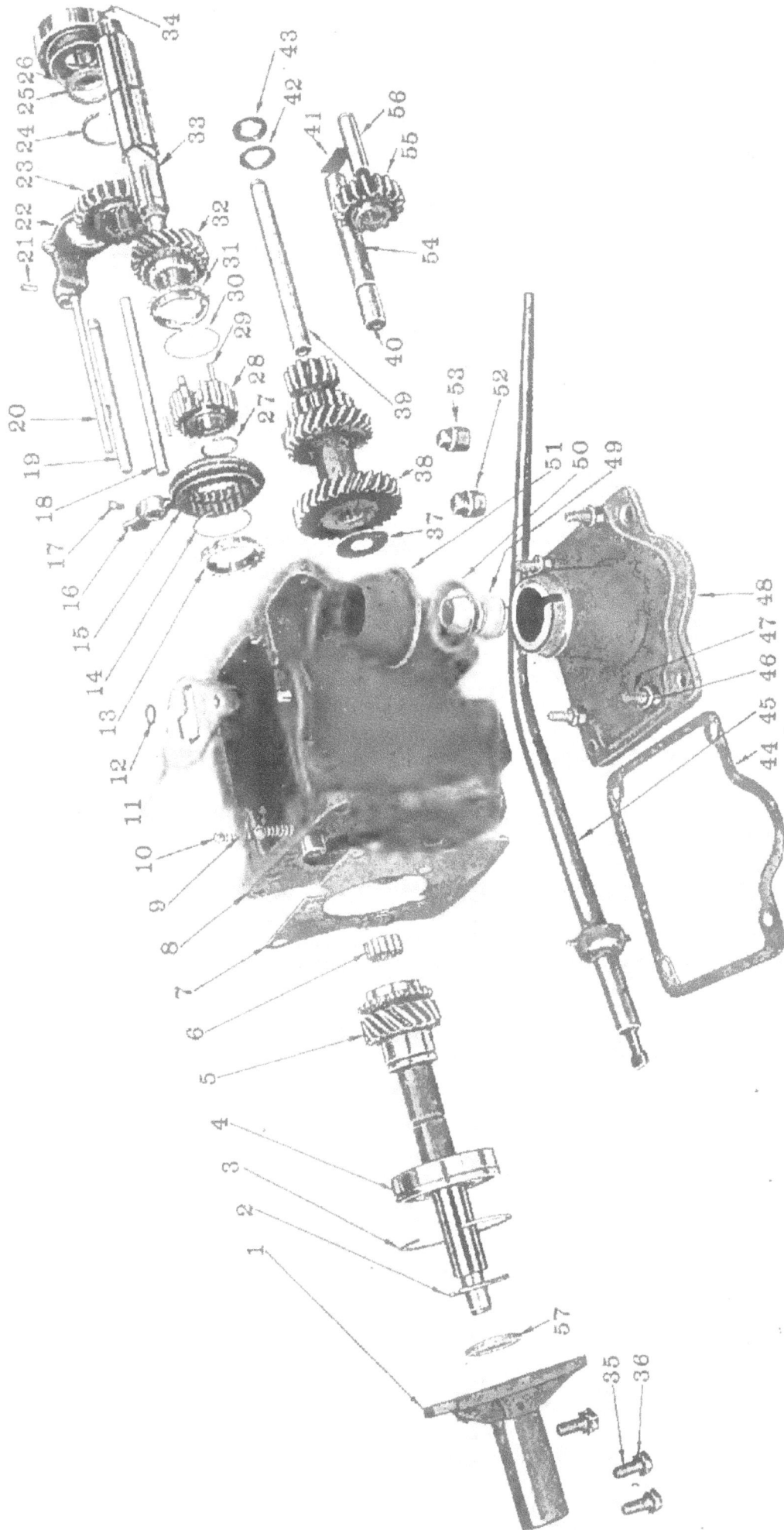

FIG. 4—TRANSMISSION

FIG. 4—TRANSMISSION (EXPLODED)

No.	Willys Part No.	Ford Part No.	Name
1	640017	GPW-7050	Transmission Main Drive Gear Bearing Retainer
2	635844	GPW-7064	Transmission Main Drive Gear Snap Ring
3	635846	B-7070	Transmission Main Drive Gear Bearing Snap Ring
4	636885	GPW-7025	Transmission Main Drive Gear Bearing
5	A-5554	GPW-7017	Transmission Main Drive Gear
6	639422	GPW-7120	Transmission Main Shaft Pilot Roller Bearing
7	637495	GPW-7051-B	Transmission Main Drive Gear Bearing Retainer Gasket
8	A-1148	GPW-7005	Transmission Case
9	635837	GPW-7234	Transmission Poppet Spring
10	635838	353081-S7	Transmission Shift Rail Poppet Ball
11	635841	GPW-7216	Transmission Shift Plate
12	635839	GPW-7208	Transmission Shift Plate Spring
13	637834	GPW-7107	Transmission Synchronizer Blocking Ring
14	637831	GPW-7109	Transmission Synchronizer Spring
15	637833	GPW-7106	Transmission Second and Direct Speed Clutch Sleeve
16	636196	GPW-7230	Transmission Shift Fork—High and Intermediate
17	636200	GPW-7245	Transmission Shift Fork Lock Screw
18	A-1155	GPW-7241	Transmission Shift Rail—High and Intermediate
19	A-1156	GPW-7240	Transmission Shift Rail—Low and Reverse
20	635836	GPW-7206	Transmission Shift Fork Guide Pin
21	636200	GPW-7245	Transmission Shift Fork Lock Screw
22	636197	GPW-7231	Transmission Shift Fork Low and Reverse
23	636879	GPW-7100	Transmission Sliding Gear—Low and Reverse
24	635844	GPW-7064	Transmission Main Shaft Snap Ring
25	A-738	GPW-7062	Transmission Main Shaft Bearing Spacer
26	A-410	GPW-7080	Transmission Oil Retaining Washer
27	637835	GPW-7059	Transmission High and Intermediate Clutch Hub Snap Ring
28	637830	GPW-7105	Transmission High and Intermediate Clutch Hub
29	637832	GPW-7116	Transmission Synchronizer Shifting Plate
30	637831	GPW-7109	Transmission Synchronizer Spring
31	637834	GPW-7107	Transmission Synchronizer Blocking Ring
32	638798	GPW-7102	Transmission Main Shaft Second Speed Gear Assembly
33	A-519	GPW-7061	Transmission Main Shaft
34	A-916	GP-7065	Transmission Main Shaft Bearing
35	635868	20366-S	Hex. Head Screw (Bearing Retainer)
36	52510	34941-S	Lockwasher
37	635812	GPW-7119	Transmission Countershaft Thrust Washer—Front
38	A-739	GPW-7113	Transmission Countershaft Gears
39	638948	GPW-7111	Transmission Countershaft
40	A-878	GPW-7121	Transmission Countershaft Gear Bushing
41	638949	GPW-7135	Transmission Countershaft and Idler Lock Plate
42	635811	GPW-7129	Transmission Countershaft Thrust Washer—Rear (Bronze)
43	A-879	GPW-7126	Transmission Countershaft Thrust Washer—Rear (Steel)
44	635861	GPW-7223	Transmission Control Housing Gasket
45	A-1380	GPW-7210	Transmission Control Lever Assembly (Gear Shift Lever)
46	635868	20366-S	Hex. Head Screw (Control Housing)
47	52045	34806-S	Lockwasher
48	635857	GPW-7204	Transmission Control Housing Assembly
49	392328	GPW-7227	Transmission Control Lever Support Spring
50	635863	BB-7228	Transmission Control Housing Cap Washer
51	A-1379	BB-7220	Transmission Control Housing Cap
52	5140	353064-S	Transmission Drain Plug
53	5140	353064-S	Transmission Filler Plug
54	A-880	GPW-7115	Transmission Countershaft Bearing Spacer
55	636882	GPW-7142	Transmission Reverse Idler Gear Assembly
56	638952	GPW-7140	Transmission Reverse Idler Gear Shaft
57	640018	GPW-7052	Front Bearing Retainer Oil Seal

TRANSMISSION SPECIFICATIONS

Transmission

Make	Warner
Model	T 84 J
Type	Synchronous Mesh
Mounting	Unit Power Plant
Shift Lever Location	On Transmission
Speeds	3 Forward—1 Reverse

Ratio

Low	2.665
Second	1.564
High	1:1
Reverse	3.554

Bearings

Clutch Shaft (Flywheel)	Bushing
Clutch Release	Ball
Clutch Shaft Rear (Main Drive Gear)	Ball
Main Shaft Front	13 Rollers
Main Shaft Rear	Ball
Counter Shaft Gear	Bushings (2)
Reverse Idler Gear	Bushing

Transmission Oil

Capacity—(Pts.)	See Capacity Chart, Page 3
S.A.E. Viscosity	See Lubrication Chart, Page 12

TRANSFER CASE

The transfer case, Fig. 1 is an auxiliary unit located at the rear of the transmission. The transfer case is essentially a two speed transmission, which provides a low gear ratio and a means of connecting the transmission to the front axle.

The shifting mechanism, Fig. 2 is located on the transfer case for engaging and disengaging the drive to the front axle, also for shifting into the low gear ratio.

On hard surface and flat roads, disengage front axle drive by placing center shift lever in forward position. The right hand lever controls the gear ratio; low and high. The low gear ratio can only be used when left hand lever is in the engaged (rear) position for front drive. Proper position for disengaging axles to use power take-off unit is shown as "N" in Fig. 4, Page 6.

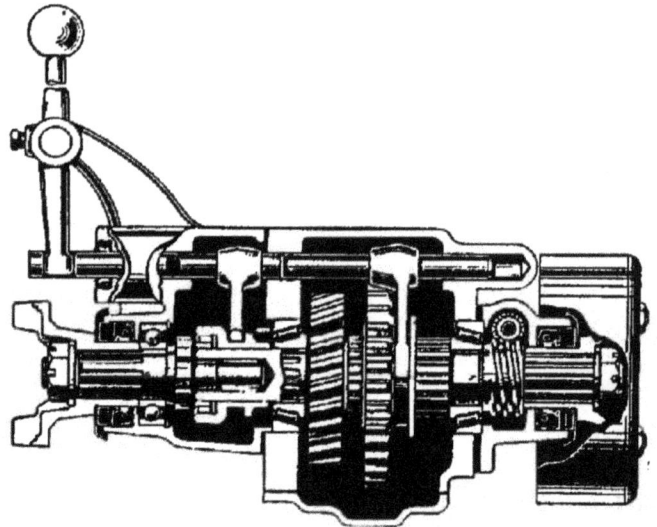

FIG. 2—TRANSFER CASE SHIFTING

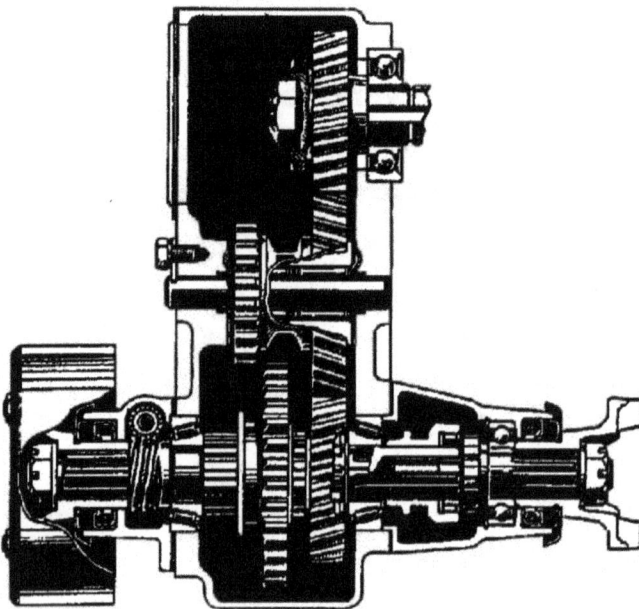

4. With a punch, drive out intermediate shaft No. 62, to rear of case.

5. Intermediate gear No. 64 with Thrust washers No. 63 and No. 65, and roller bearings No. 61 and 66 can be removed through the bottom of case.

6. Remove poppet plugs No. 18, springs, No. 17, and balls No. 16 on both sides of output bearing cap, No. 14. Shift front wheel drive to engaged position (shaft forward).

7. Remove cap screws holding front output bearing cap No. 14 and remove cap as an assembly, with universal joint end yoke No. 26, clutch shaft No. 12, bearing No. 10, clutch gear No. 3, fork No. 2, and shift rod No. 8. Taking care not to lose inter-lock No. 7.

FIG. 1—TRANSFER CASE

8. Remove output shaft snap ring No. 70 and thrust washer No. 69.

9. Remove cap screws holding rear output bearing cap No. 43, and remove cap as an assembly with universal joint flange No. 33, speedometer gears No. 40 and 45, bearing No. 51, and output shaft No. 67. This will allow sliding gear No. 56 and output shaft gear No. 68 to slide off the output shaft No. 67 and come out through the bottom of case.

Disassembling Transfer Case

To remove the gears and bearings from the transfer case on the bench, the following procedure is recommended.

1. Remove brake band assembly and linkage.

2. Remove cap screws and lockwashers holding lower cover, No. 55, Fig. 3.

3. Remove lock plate screw, lockwasher and lockplate, No. 50.

10. Remove set screw No. 48 in sliding gear shift fork No. 49. This will allow shift rod No. 9 to slide through the fork and hole in case. The fork can then be removed through bottom of case.

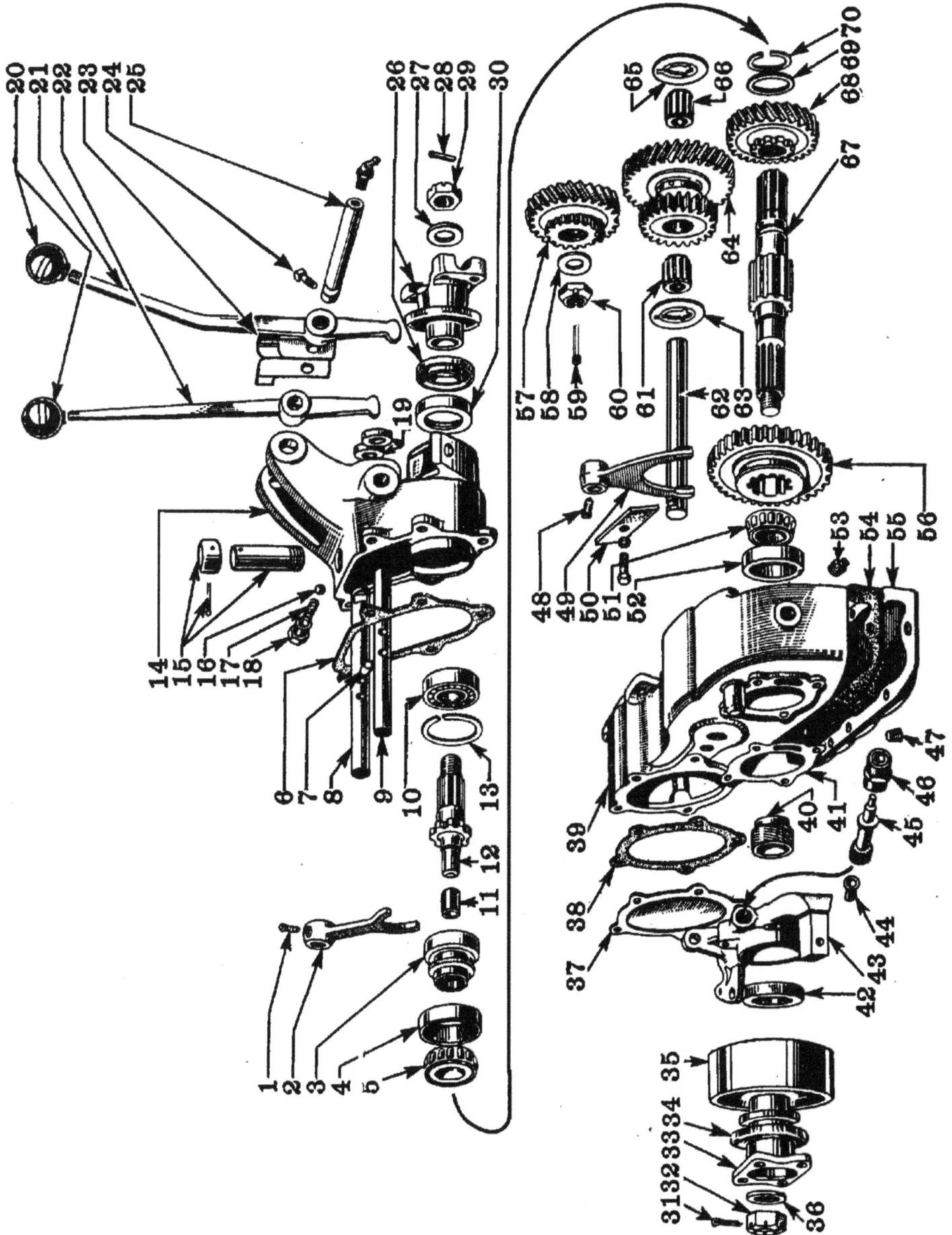

FIG. 3—TRANSFER CASE EXPLODED VIEW

FIG. 3—TRANSFER CASE EXPLODED VIEW

No.	Willys Part No.	Ford Part No.	Name
1	A-963	355550-S	Shift Fork to Rod Set Screw
2	A-960	GP-7711	Front Wheel Drive Shift Fork
3	A-992	GP-7762	Output Shaft Clutch Gear
4	52883	O1V-1202	Output Shaft Bearing Cup
5	51575	GP-7723	Output Shaft Bearing Cone and Rollers
6	A-957	GPW-7773	Output Shaft Bearing Cap Gasket—Front
7	A-965	GP-7789	Shift Rod Interlock
8	A-962	GP-7787	Front Wheel Drive Shift Rod
9	A-1504	GPW-7786	Under Drive Shift Rod
10	A-1007	GP-7719	Output Clutch Shaft Bearing
11	A-987	GP-7777	Output Clutch Shaft Pilot Bushing
12	A-976	GP-7761	Output Clutch Shaft
13	A-978	GP-7783	Output Clutch Shaft Bearing Snap Ring
14	A-968	GPW-7774	Output Shaft Bearing Cap—Front
15	A-934	GP-7754	Transfer Case Breather Assembly
16	5599	353075-S	Shift Rod Poppet Ball
17	A-966	GP-7788	Shift Rod Poppet Spring
18	A-967		Shift Rod Poppet Plug
19	A-974	GP-7708	Shift Rod Oil Seal
20	A-971	GP-7213	Shift Lever Handle
21	A-1505	GPW-7793	Under Drive Shift Lever
22	A-1506	GPW-7710	Front Wheel Drive Shift Lever
23	A-970	GP-7799	Shift Lever Spring
24	A-973	355378-S	Shift Lever Pivot Pin Set Screw
25	A-972	GP-7796	Shift Lever Pivot Pin
26	A-1106	GP-7729	Universal Joint End Yoke—Front
27	A-1028	356504-S	Companion Flange Nut Washer
28	5108	72053-S	Companion Flange Nut Cotter Pin
29	A-980	356125-S	Companion Flange Nut
30	A-958	GP-7770-A	Output Shaft Oil Seal
31	5108	72053-S	Companion Flange Cotter Pin
32	A-980	356125-S	Companion Flange Nut
33	A-1105	GP-4863	Companion Flange—Rear
34	A-1111	GP-7776	Dust Shield
35	A-1002	GP-2614	Brake Drum
36	A-1028	356504-S	Companion Flange Washer
37	A-1508	GPW-7706	Rear Cover
38	A-1509	GPW-7707	Rear Cover Gasket
39	A-1503	GPW-7705	Transfer Case
40	A-1511	GP-17285	Speedometer Drive Gear
41	A-982	GP-7782-A	Output Shaft Bearing Shim
42	A-958	GP-7770-A	Output Shaft Oil Seal
43	A-1507	GPW-7769	Output Shaft Bearing Cap—Rear
44	A-985	GP-17277	Speedometer Drive Pinion Bushing
45	A-1512	GPW-17371	Speedometer Driven Gear
46	635396	GP-17333	Speedometer Driven Gear Sleeve
47	A-1104	358059-S	Transfer Case Drain Plug
48	A-963	355550-S	Shifting Fork to Rod Set Screw
49	A-959	GP-7712	Under Drive Shift Fork
50	A-1001	GP-7707	Intermediate Shaft Lock Plate
51	51575	GP-7723	Output Shaft Bearing Cone and Rollers
52	52883	O1V-1202	Output Shaft Bearing Cup
53	5140	353064-S	Filler Plug
54	A-954	GP-7709	Transfer Case Cover Gasket—Bottom
55	A-953	GP-7708	Transfer Case Cover—Bottom
56	A-988	GP-7765	Output Shaft Sliding Gear
57	A-1510	GP-7722	Main Shaft Gear
58	A-1410	356560-S	Main Shaft Washer
59	5397	72071-S	Main Shaft Nut Cotter Pin
60	A-520	356134-S18	Main Shaft Nut
61	A-924	GP-7718	Intermediate Gear Bearing
62	A-998	GP-7743	Intermediate Shaft
63	A-1000	GP-7744	Intermediate Gear Thrust Washer
64	A-999	GP-7742	Intermediate Gear
65	A-1000	GP-7744	Intermediate Gear Thrust Washer
66	A-924	GP-7718	Intermediate Gear Bearing
67	A-1764		Output Shaft
68	A-989	GP-7766	Output Shaft Gear
69	A-990	GP-7771	Output Shaft Gear Thrust Washer
70	A-991	GP-7784	Output Shaft Gear Snap Ring

Disassembly of Front Cap Assembly

1. Remove cotter pin, No. 28, nut, No. 29 and washer No. 27.
2. Remove universal joint yoke No. 26.
3. Remove oil seal No. 30.
4. Remove set screw No. 1 and shifting rod No. 8.
5. Clutch gear No. 3 and fork No. 2 can now be removed together.
6. Remove output clutch shaft No. 12 carefully pressing through the bearing No. 10.
7. Remove snap ring No. 13.
8. Remove bearing No. 10.

Disassembly of Rear Cap Assembly

1. Remove cotter pin No. 31, nut, No. 32, and washer No. 36.
2. Remove companion flange No. 33.
3. Remove oil seal No. 42.
4. Remove speedometer driven gear, No. 45.
5. Output shaft No. 67 can now be removed from cap No. 43 after which bearing cone No. 51 and the speedometer driving gear No. 40 can be pressed off the shaft.

Shims No. 41 provided between the rear cap No. 43 and case No. 34 are for adjustment of the roller bearings No. 51 and 5. Bearings should be adjusted so that there is not more than .003" end movement of the shaft No. 67.

Reassembling is merely a reversal of the foregoing. When assembling the transfer case to the transmission be sure that the countershaft lock plate No. 41, Fig. 4, Transmission Section is properly located between the two shafts and fits into transfer case. Make certain that all parts are carefully washed and free from all dirt and foreign matter.

Assembly of Transmission and Transfer Case to Vehicle

The installation of the assembly to the engine is the reversal in the operations for disassembly as covered under heading "Removal of Transmission and Transfer Case". For illustration of snubbing rubber and rear engine mounting see Fig. 4.

After assembling to the engine be sure that the clutch pedal has ¾" free pedal travel; refer to Page 37 for "Clutch Pedal Adjustment." Fill both the transmission and the transfer cases with the proper lubricant. See Lubrication Chart, Page 12.

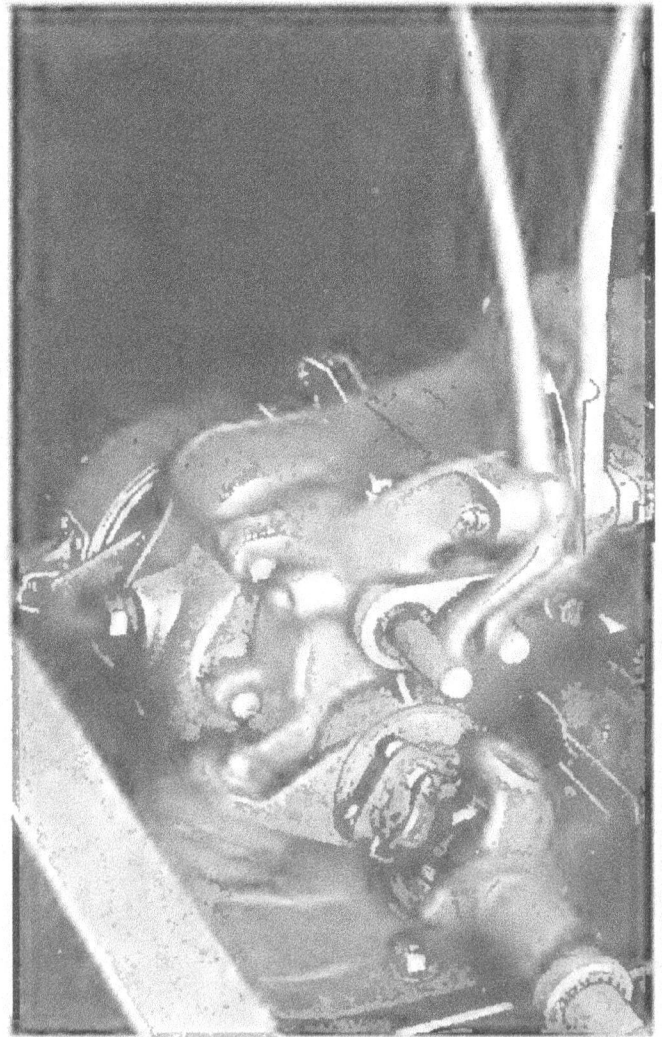

FIG. 4—TRANSFER CASE SNUBBING RUBBER

TRANSFER CASE TROUBLES AND REMEDIES

SYMPTOMS	PROBABLE REMEDY

Slips Out of Gear (High-Low)

Shifting Lock Spring Weak	Replace spring
Bearing Broken or Worn	Replace
Shifting Fork Bent	Replace

Slips Out Front Wheel Drive

Shifting Lock Spring Weak	Replace
Bearing Worn or Broken	Replace
End Play in Shaft	Adjust (See Instructions)
Shifting Fork Bent	Replace

Hard Shifting

Lack Lubrication	Drain and Refill—3 pints
Shift Lever Stuck on Shaft	Remove, clean-lubricate
Shifting Lock Ball Scored	Replace Ball
Shifting Fork Bent	Replace Fork
Low Tire Pressure	Inflate all tires—30 lbs.

Grease Leak at Front or Rear Drive

Grease leak at covers	Install new Gaskets
Grease leak between Trans. and Transfer cases	Install new Gaskets
Grease leak at Output Shafts	Install new oil Seal

TRANSFER CASE SPECIFICATIONS

Transfer Case

Make	Spicer
Model	18
Mounting	Unit with Transmission
Shift Lever	Floor
Ratio	High 1:1 / Low 1.97:1

Transfer Case Bearings

Transmission Mainshaft	Ball
Idler Gear	2 Roller
Out Put Shaft	Taper Roller
Front Axle Clutch Shaft	
Front Bearing	Ball
Rear Pilot in Output Shaft	Bronze Bushing I.D. .627"

Transfer Case Oil

Capacity Pts.	See Capacity Chart, Page 3
S.A.E. Viscosity	See Lubrication Chart, Page 12

Speedometer Drive

Drive Gear Teeth	4
Driven Gear Teeth	14

PROPELLER SHAFTS AND UNIVERSAL JOINTS

The drive from the transmission to the front and rear axles is accomplished through a propeller shaft and two universal joints. Fig. 1.

The splined slip joint at one end of each shaft allows for variations in distance between the transfer case and the front and rear axle units due to spring action.

The slip joint is marked with arrows at the spline and the sleeve yoke. Note markings to facilitate proper assembly so the yokes of the universal joints at front and rear of shaft are in the same plane when assembled, Fig. 2.

The propeller shaft connecting the transfer case with the front axle has the "U" bolt type universal joint at both ends.

The rear propeller shaft is equipped with the "U" bolt type joint at the rear where it attaches to the rear axle. The front universal joint is the snap ring type.

These universal joints are the Needle Bearing type and are so designed that correct assembly is a very simple matter. No hand fitting or special tools are required.

The journal trunnion and needle bearing assemblies are the only parts subject to wear, and when it becomes necessary to replace these parts, the propeller shaft should be removed from the vehicle.

FIG. 1—PROPELLER SHAFT ASSEMBLY

Disassembling of Snap Ring Universal Joints

To remove snap rings pinch ends together with a pair of pliers. If the ring does not readily snap out of the groove tap the end of the bearing lightly, this will relieve pressure against the ring. See Fig. 3.

Drive on the end of one bearing until the opposite bearing is pushed out of the yoke. Turn the joint over and drive the first bearing back out of its lug by driving on the exposed end of the journal shaft. Use a soft round drift with a flat face about 1/32″ smaller in diameter than the hole in the yoke, otherwise there is danger of damaging the bearing.

Repeat this operation for the other two bearings, then lift out journal assembly, sliding to one side and tilting over the top of the yoke lug.

Wash all parts in cleaning solution and if parts are not worn, lubricate with a good grade of semi-fluid lubricant, see Lubrication Chart, Page 12. Make sure the reservoir in each journal trunnion is filled. Put the rollers in the race and fill the race about one-third full. It is advisable to install new gaskets, No. 2, Fig. 4 on the journal assembly.

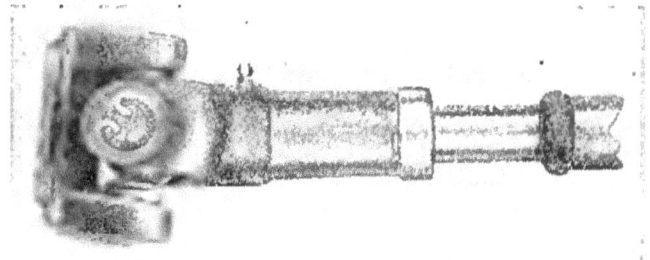

FIG. 2—ARROW MARKING

Reassembling of Snap Ring Universal Joints

Reassembling is merely a reversal of the dismantling operation. Hold the bearing in a vertical position to prevent needles from dropping out of bearing race when installing in joint.

When assembled, if joint appears to bind tap the lugs lightly with a hammer which will relieve any pressure on the bearings at the end of the journal.

When inserting the spline of the propeller shaft into the universal joint be sure that the arrows on the propeller shaft and yoke sleeve are in line. See Fig. 2.

Disassembly of "U" Bolt Type Universal Joint

Removal of the "U" Bolts at axle and transmission end yoke allows the complete propeller shaft assembly to be removed.

After removing "U" bolt slide sleeve yoke (slip joint) towards the shaft which will allow the bearing race to come out from behind the shoulders on end yoke. Care should be taken to hold bearing races in place to avoid losing the rollers.

Now remove snap lock ring, No. 1, Fig. 4 in the sleeve yoke at front and stud ball yoke at rear end of shaft by pinching ends together with a pair of pliers. If a ring does not snap readily out of the groove, tap the end of the bearing lightly, which will relieve the pressure against the ring.

Drive on the end of one bearing until the opposite bearing is pushed out of the yoke. Turn the universal joint over and drive the first bearing out by driving on the exposed end of the journal assembly. Use a soft round drift with a flat face about 1/32″ smaller in diameter than the hole in the yoke, otherwise there is danger of damaging the bearing.

Now lift out journal assembly by sliding to one side and tilting over the top of the yoke lug. Clean all parts and if parts are not worn, repack with a good grade of semi-fluid lubricant, see Lubrication Chart, Page 12. Make sure the reservoir in the end of each trunnion is filled. With the rollers in the race, fill the race about one-third full. It is advisable to install new gaskets on journal assembly.

Reassembling of "U" Bolt Type Universal Joint

Reassembling is merely a reversal of the dismantling operation.

Be sure to hold the bearing in a vertical position to prevent the needles from dropping out of the bearing race.

When assembled, if joints appear to bind tap the lugs lightly with a hammer which will relieve any pressure on the bearings at the end of the journal.

When assembling the bearings into the end yoke the use of a "C" Clamp over the extreme ends of the bearing races to draw the bearings into correct position will greatly facilitate seating them inside of the bearing shoulders on the end yokes. "U" bolt, torque wrench reading, 15-18 ft. lbs.

When inserting the propeller shaft spline into the universal joint be sure that the arrows on the propeller shaft and yoke sleeve are in line. See Fig. 2.

Lubrication

Do not use grease in the needle bearings.

At each 1,000 mile lubrication job, lubricate the Universal Joints, using a hand gun. See Lubrication Chart for oil specifications.

The sliding spline shaft should be lubricated with a good grade of grease or oiled every 1,000 miles, or every time the chassis is lubricated. A hydraulic pressure fitting is provided for this purpose on the side of the sleeve yoke.

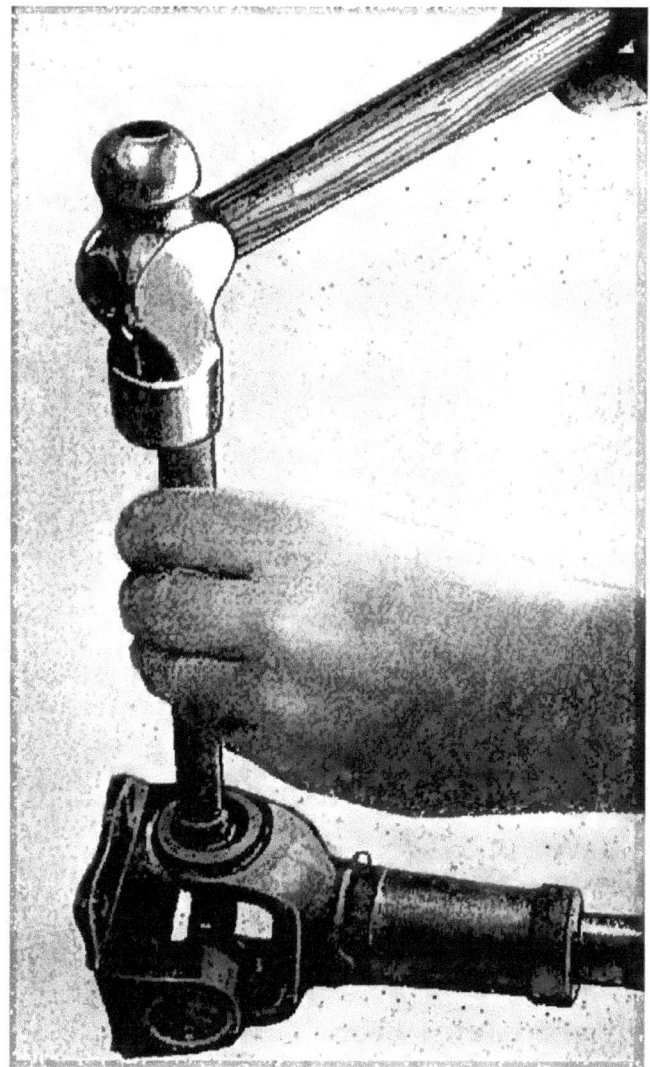

FIG. 3—REMOVING UNIVERSAL JOINT BEARING

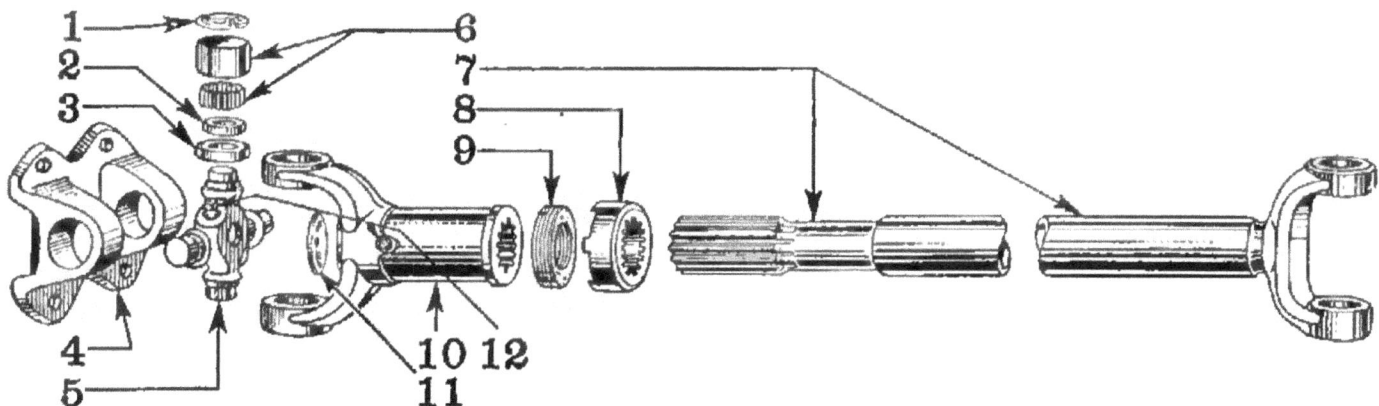

FIG. 4—PROPELLER SHAFT—REAR

No.	Willys Part No.	Ford Part No.	Name	No.	Willys Part No.	Ford Part No.	Name
1	A-945	O1Y-7096	Universal Joint Bearing Snap Ring	7	A-1429	GPW-4605	Propeller Shaft Tube Assembly—Rear
2	A-941	O1T-7078-A	Trunnion Gasket	8	A-942	GP-7077	Dust Cap
3	A-940	O1Y-7083	Trunnion Gasket Retainer	9	A-943	GP-7097	Cork Washer
4	A-950	GP-4866	Universal Joint Flange Yoke	10	A-935	GP-7092	Universal Joint Sleeve Yoke Assembly
5	A-1426	GPW-7084	Universal Joint Journal Assembly	11	A-937		Sleeve Yoke Plug
6	A-1425	GPW-7099	Universal Joint Bearing Race	12	638792	353043-S7	Hydraulic Fitting

PROPELLER SHAFT AND UNIVERSAL JOINT SPECIFICATIONS

Propeller Shaft

Make.................................. Spicer
Shaft Diameter....................... 1¼"
Length (Front)........(Joint center to center) 21¹¹⁄₁₆"
Length (Rear)........(Joint center to center) 20¹⁄₃₂"

Universal Joint Front Drive................. Front

Make.................................. Spicer
Type................................. U Bolt and Snap Ring
Model................................ 1268
Bearings............................. Needle Roller Spicer 98-851

Universal Joint Front Drive................. Rear

Make.................................. Spicer
Type................................. Snap Ring and U Bolt
Model................................ 1261
Bearings............................. Needle Roller Spicer 98-851

Universal Joint Rear Drive................. Front

Make.................................. Spicer
Type................................. Snap Ring Slip Joint
Model................................ 1261
Bearings............................. Needle Roller Spicer 98-851

Universal Joint Rear Drive................. Rear

Make.................................. Spicer
Type................................. U Bolt and Snap Rings
Model................................ 1268
Bearing.............................. Needle Roller Spicer 98-851

Lubricant............................. See Lubrication Chart, Page 12

FRONT AXLE

The front axle assembly is a front wheel driving unit with specially designed steering knuckles, Fig. 1 and a conventional type differential with hypoid drive gears.

The front wheels are driven by axle shafts equipped with constant velocity universal joints which are enclosed in the steering knuckle housing.

The differential is mounted in the housing similar to that used in rear axle, except that the drive pinion shaft is toward the rear instead of the front and to the right of the center of the axle. This design allows placing front propeller shaft along right side of engine oil pan without reducing road clearance under engine.

The differential parts are interchangeable with those of the rear axle.

The axle is the full-floating type and can be removed without disassembling the steering knuckle.

Axle Shaft and Universal Joint Assembly

To remove axle shaft and universal joint assembly the following operations should be performed. See Fig. 2.

1. Remove wheel assembly.
2. Remove hub cap by inserting two screw drivers from opposite sides behind inner flange on cap and pry off.
3. Remove axle shaft cotter pin, nut and washer.
4. Remove axle shaft drive flange bolts and lock washers.
5. Apply the foot brakes and remove the axle shaft flange with puller furnished in tool kit, see Fig. 3.
6. Remove wheel bearing nuts and washers, No. 22, Fig. 2. First bend the lip on lock washer, No. 21 away from the nut with a chisel, remove the outer nut, lock washer, adjusting nut and bearing lock washer, Fig. 4. Wrench is furnished in tool kit.

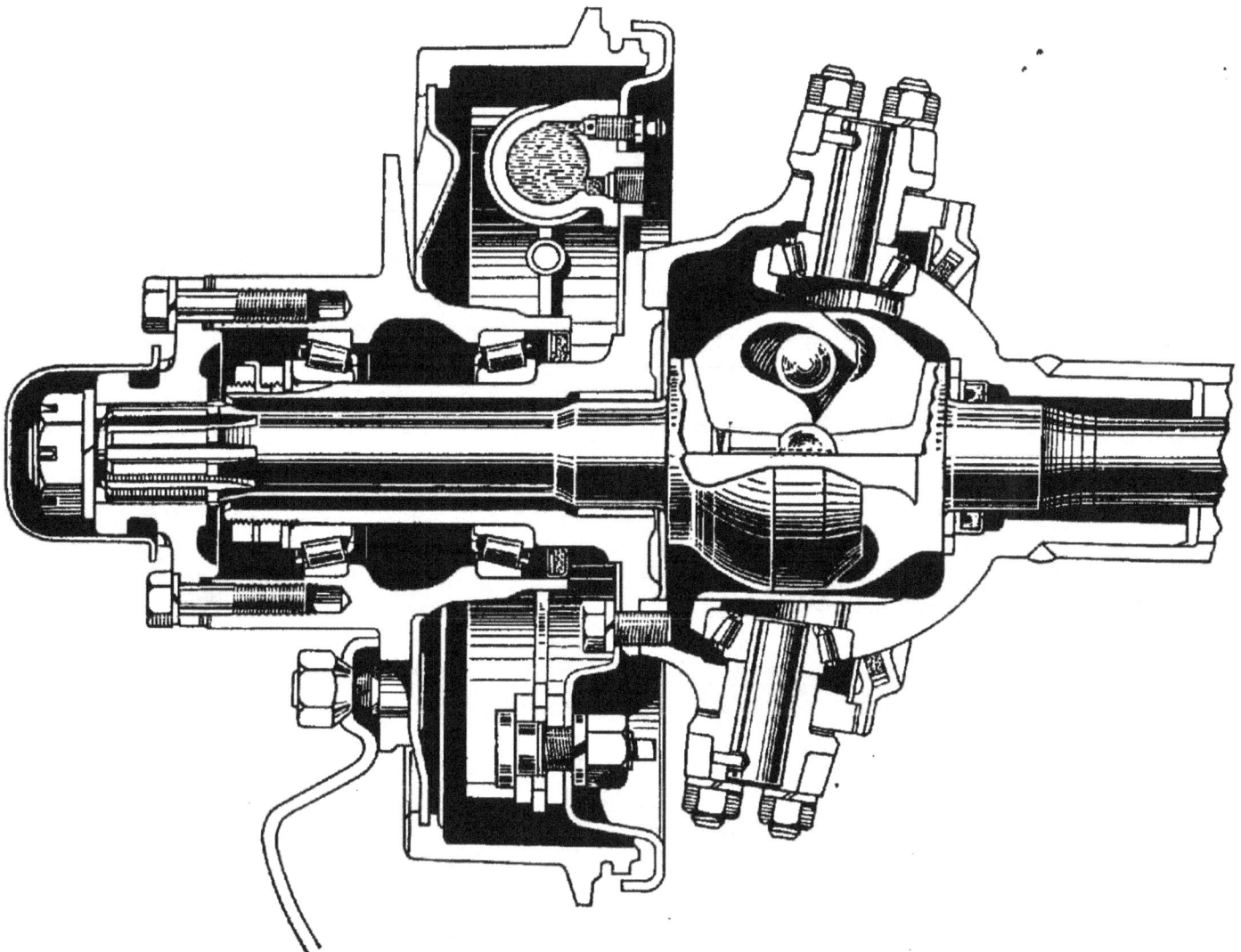

FIG. 1—FRONT WHEEL

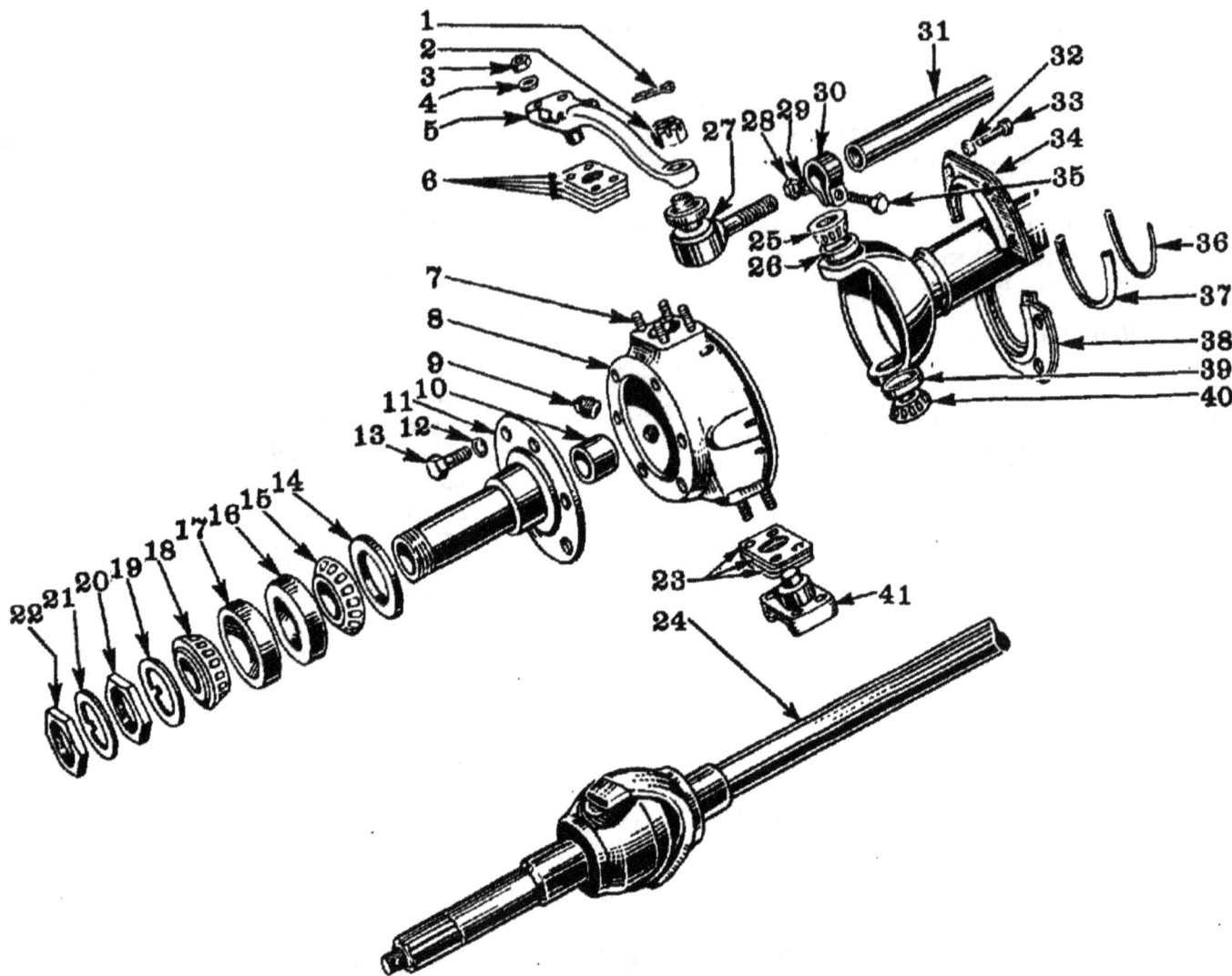

FIG. 2—FRONT AXLE, STEERING KNUCKLE AND WHEEL BEARINGS

(Bendix Universal Joint)

No.	Willys Part No.	Ford Part No.	Name	No.	Willys Part No.	Ford Part No.	Name
1	5152	72025-S	Tie Rod Stud Nut Cotter Pin	24	A-809	GPW-3206-A	Axle Shaft and Universal Joint Assembly (Bendix type) — Right Hand (Ford GPW-3207-A; Willys A-810 Left Hand)
2	10558	351059-S7	Tie Rod Stud Nut				
3	630598		Steering Arm Nut				
4	5010	34807-S	Steering Arm Nut Lockwasher	25	52940	GP-3161	King Pin Bearing Cone and Rollers
5	A-1712	GPW-3113	Upper Steering Arm—Left Hand (Ford GPW-3112; Willys A-1710 Right Hand)	26	52941	GP-3162	King Pin Bearing Cup
				27	A-847	GP-3290	Tie Rod Socket Assembly Left Hand (Ford GP-3289; Willys A-838 Right Hand)
6	A-830	GP-3117-A	King Pin Adjusting Shims				
7	A-1714	357703-S	Steering Arm Stud—Upper (A-5504 Dowel Stud—Upper Outside Front and Inside Rear)	28	636575	34083-S2	Tie Rod Socket Clamp Nut
				29	5010	34807-S	Tie Rod Socket Clamp Nut Lockwasher
8	A-811	GP-3148-A2	Steering Knuckle Right Hand (Ford GP-3149-A2; Willys A-812 Left Hand)	30	A-1706	51-3287	Tie Rod Socket Clamp
				31	A-1705	GPW-3281	Tie Rod Tube Right Hand (Ford GPW-3282; Willys A-1709 Left Hand)
9	5140	353064-S	Steering Knuckle Filler Plug				
10	A-853	GP-3205	Wheel Bearing Spindle Bushing	32	52510	34941-S	Knuckle Oil Seal Screw Lockwasher
11	A-851	GP-3105	Wheel Bearing Spindle Assembly	33	A-872	355483-S	Knuckle Oil Seal Screw
12	5010	34807-S	Brake Disc Screw Lockwasher	34	A-813		Steering Knuckle Oil Seal Assembly—Half
13	A-877	355552-S	Brake Disc Screw				
14	A-864	GP-1177	Hub Oil Seal Assembly	35	A-1707	24916-S2	Tie Rod Socket Clamp Screw
15	52942	GP-1201	Wheel Bearing Cone and Rollers	36	A-918	GP-3139	Steering Knuckle Oil Seal Felt Pressure Strip
16	52943	GP-1202	Wheel Bearing Cup				
17	52943	GP-1202	Wheel Bearing Cup	37	A-819	GP-3135	Steering Knuckle Oil Seal Felt—Half
18	52942	GP-1201	Wheel Bearing Cone and Rollers	38	A-813	GPW-1088	Steering Knuckle Oil Seal Assembly—Half
19	A-865	GP-1218	Wheel Bearing Lockwasher				
20	A-866	GP-4252	Wheel Bearing Nut	39	52941	GP-3162	King Pin Bearing Cup
21	A-867	GP-1124	Wheel Bearing Nut Lockwasher	40	52940	GP-3161	King Pin Bearing Cone and Rollers
22	A-866	GP-4252	Wheel Bearing Nut	41	A-828	GP-3140	Lower King Pin Bearing Cap
23	A-830	GP-3117-A	King Pin Adjusting Shims				

7. Remove wheel hub and drum assembly with bearings taking care not to damage oil seal.
8. Remove brake tube and brake backing plate screws, No. 13, Fig. 2.
9. Remove spindle No. 11.
10. The complete axle shaft and universal joint assembly No. 24 can now be pulled out of the axle housing. Care should be taken not to injure the outer oil seal assembly in axle housing.

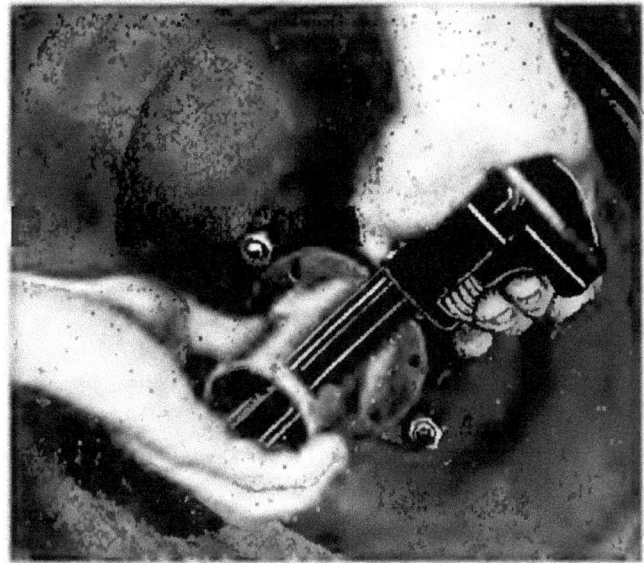

FIG. 4—REPLACING HUB NUT

Inspect the ball raceways for excessive wear. Fig. 6. If a raceway is badly worn the complete axle and universal joint assembly should be replaced. If the center ball pin is worn, it should be replaced. Inspect the center ball and the four driving balls for scratches, grooves or flat spots and replace if necessary. The driving balls (.875" diameter) are available from .003" undersize to .003" oversize in steps of .001" to permit selective fitting. If any or all of the driving balls are to be replaced the old ball or balls should be measured with a micrometer and the same size new balls used. Selective assembly is not required when installing a new center ball.

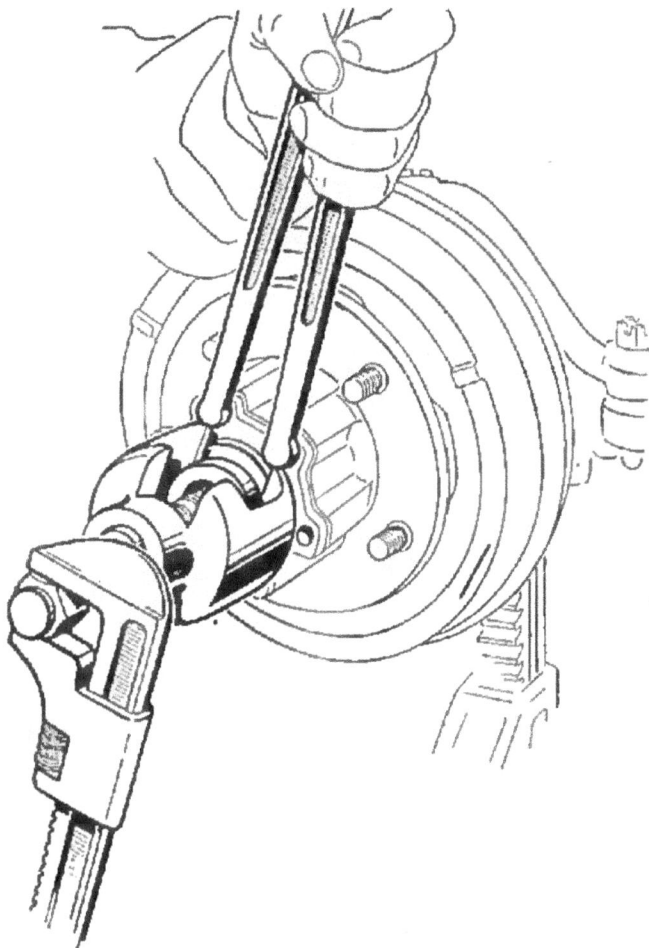

FIG. 3—PULLING DRIVING FLANGE

Disassembly (Bendix Joint)

After the axle shaft assembly has been removed, the universal joint may be disassembled as follows:

1. Wash the axle shaft and universal joint thoroughly in cleaning fluid.
2. Using a drift and hammer, drive out the retainer pin which locks the center ball pin in wheel end of shaft. See Fig. 5.
3. Bounce the wheel end of the shaft on a block of wood to cause the center ball pin to move into the drilled passage in the wheel end of the shaft.
4. Pull the two halves of the joint apart and then bend sharply at the universal. Rotate the center ball until grooved side lines up with ball raceway. This permits the adjacent ball to be moved past the center ball and removed from the joint. The remaining three driving balls and center ball will then drop out.

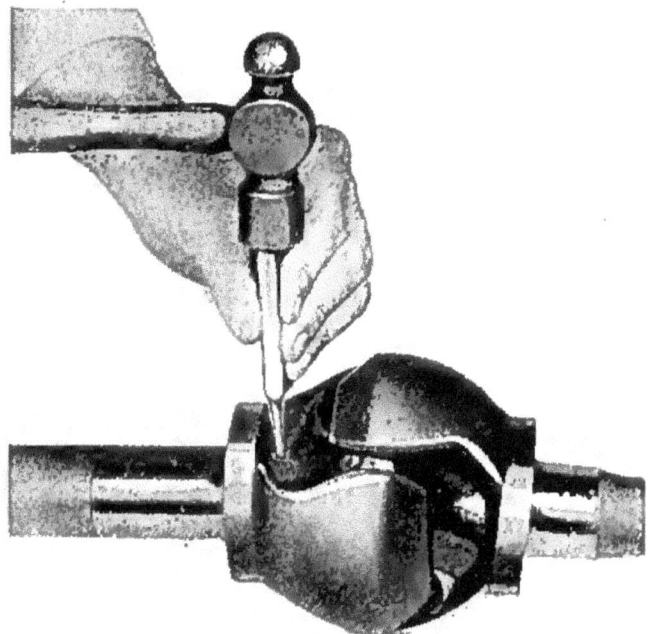

FIG. 5—REMOVING RETAINER PIN
(Bendix Joint)

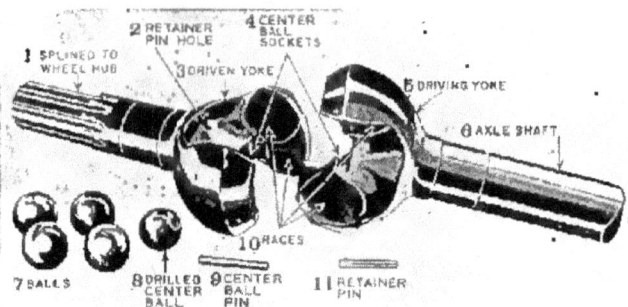

FIG. 6—AXLE SHAFT UNIVERSAL JOINT

Reassembly—(Bendix Joint)

1. Place the differential half of the axle shaft in a bench vise, with the ground portion of the shaft above the vise jaws.
2. Install the center ball (one with hole drilled in it) in its socket in the shaft, hole and groove facing you.
3. Drop the center ball pin into the drilled passage in the wheel half of the shaft.
4. Place the wheel half of the shaft on the center ball. Then slip three balls into the raceways.
5. Turn the center ball until the groove in it lines up with the raceway for the remaining ball as shown in Fig. 7. Slip the ball into the raceway and straighten up the wheel end of the shaft.
6. Turn the center ball until the center ball pin drops into the hole drilled in the ball.
7. Install the retainer pin (lock pin) and prick punch both ends to securely lock in place.

Disassembly (Rzeppa Joint)

After the shaft has been removed, the universal joint may be disassembled as follows, Fig. 4, Pg. 115:

1. Remove the three screws holding the front axle shaft to the joint and pull the shaft out of the splined inner race. To remove the axle shaft retainer, remove the retainer ring on the axle shaft.
2. Clean the universal joint in a suitable cleaning solution and lift out the axle centering pin.
3. Push down on various points of the inner race and cage until the balls can be taken out with the help of a small screw driver. Be careful not to damage parts.
4. After all the balls have been removed the inner race and cage can be turned over so the pilot cup is up, then remove the pilot cup.
5. There are two large elongated holes in the cage as well as four small holes. Turn the cage so

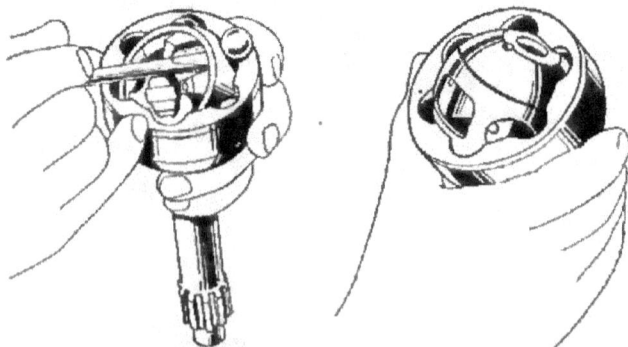

FIG. 7—ASSEMBLING UNIVERSAL JOINT BALLS

two bosses in the spindle shaft will drop into the elongated holes and lift out cage.

6. To remove the inner race turn it so one of the bosses will drop into an elongated hole in the cage, shift the race to one side, and lift out opposite side.

Reassembly (Rzeppa Joint)

1. Reassembly of the joint is the reverse of dismantling. Care should be exercised not to damage parts and see that they are clean of all dirt and grit.

To Reassemble Axle Shaft and Universal Joint Assembly to Housing

1. Clean all parts so that they are free from dust and foreign matter.
2. Enter universal joint and axle shaft assembly in the housing, taking care not to injure the outer and inner oil seals. Enter spline end of axle into the differential and push in until the shoulder on the universal joint stops against the axle.
3. Install wheel bearing spindle.

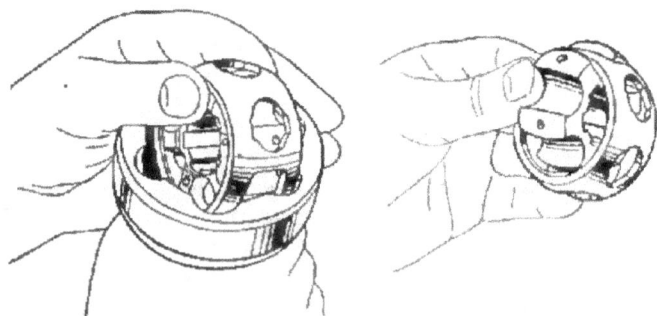

FIG. 8—DISMANTLING RZEPPA JOINT

FIG. 9—REMOVING CAGE—RZEPPA JOINT

4. Install brake tube and bolt backing plate in position.

5. Grease wheel bearings and assemble bearings, wheel hub and drum on the wheel bearing spindle. Install wheel bearing washer, No. 19, Fig. 2, and adjusting nut, No. 20. Tighten nut until there is a slight drag on the bearings, when the wheel is turned, then back off approximately one-quarter turn. Install lock washer No. 21 and nut No. 22, tightening nut into place and then bending the lock washer over on the lock nut.

6. With Bendix joint install axle drive flange on axle splines, without shims.

Measure the space with a Feeler Gauge between the outer end of the wheel hub and the inner face of the drive flange, Fig. 10. This will determine the amount of shims to be installed. In order to have proper clearance in the universal joint, it is necessary to add a .040" shim to those required as measured by the Feeler Gauge.

Remove driving hub and install the correct amount of shims replacing driving hub on spline shaft and install six cap screws.

With Rzeppa joint be sure to install all the shims as removed when dismantling the axle drive flange. (.060" shims in each side).

7. Assemble axle shaft washer, nut and cotter pin.

8. Install the Hub Cap.

9. Assemble Wheel.

10. Check front wheel alignment, which is covered under "Steering".

11. Bleed Brake.

Make certain the steering knuckle universal joint is lubricated through the filler plug in the knuckle housing. See Lubrication Chart, Page 12.

FIG. 10—CHECKING FLANGE END PLAY

Replacing Steering Knuckle Bearing

Replacement of the bearings or bearing cups on the king pins necessitates removal of the hub and brake drum assembly, wheel bearings, axle shaft, wheel bearing spindle and the steering knuckle. The steering knuckle should be disassembled as follows:

1. Remove the eight screws No. 33, Fig. 2 which hold the oil seal retainers in place No. 34 and 38.

2. Remove the four nuts holding the lower king pin bearing cap, No. 41.

3. Remove the four nuts, No. 3 holding the upper steering arm in place, and remove brake hose shield also arm No. 5. The steering knuckle No. 8 can now be removed from the axle.

4. Wash all parts in cleaning solution and inspect bearings and races for scores, cracks or chips. All damaged parts should, of course, be replaced.

In the event the bearing cups are damaged, they can be removed by the use of a driver or a suitable drift.

Reassembling Steering Knuckle

Reverse the procedure outlined above to reassemble the unit. When reinstalling the steering knuckle, sufficient shims must be installed under the arm and lower bearing cap so the proper tension will be maintained on the bearing. The shims are available in thicknesses of .003", .005", .010" and .030".

Install one each of the .003", .005", .010" and .030" shims over studs on the steering knuckle, top and bottom. Install the arm, and lower bearing cap, lock washers, and nuts, and tighten securely. Check the tension of the bearings by hooking checking scale in the hole in the arm at tie rod and socket, either remove or add shims until the load is approximately 25 to 35 inch pounds, without oil seal assembly in position. Make sure there are the same thickness of shims between arm and knuckle as between lower cap and knuckle.

Steering Knuckle Oil Seal

Replacement of the oil seal No. 34 and 38 can be made very easily be merely removing the eight screws which hold the oil seal in place. Before reinstalling the oil seal examine the spherical surface of the axle for scores or scratches which might damage the seal. Roughness of any kind should be smoothed down with emery cloth.

Reinstall both upper and lower halves of the oil seal, making sure that the felt fits snugly at the points where the upper and lower halves come together, Fig. 11.

After driving in wet, freezing weather swing the

FIG. 11—STEERING KNUCKLE OIL SEAL

front wheels from right to left to remove moisture adhering to the oil seal and the spherical surface of the universal joint housing. This will prevent freezing with resulting damage to oil seal felts. Should the car be stored for any period of time, coat these surfaces with light grease to prevent rusting.

Axle Shaft Outer Oil Seal

In the event it should be necessary to replace the axle shaft outer oil seal, remove the axle shaft and universal joint as described under the "Axle Shaft and Universal Joint".

The oil seal is a light press fit in the housing

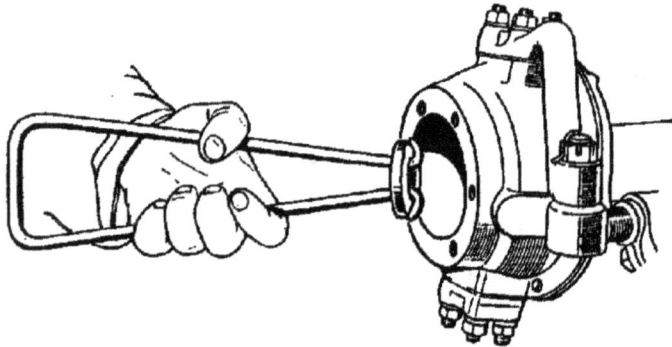

FIG. 12—REMOVING OIL SEAL

and will require a tool or puller for removal. Insert the ends of the puller behind the oil seal and tap the end of the puller lightly with a hammer. See Fig. 12.

Before installing a new seal make sure it has been soaked thoroughly in oil. This will not only make the leather more pliable but will avoid it being burned by friction with the axle shaft when the vehicle is driven.

After placing the oil seal in position in the housing, it can easily be driven in place by using a driver or a block of hard wood and a hammer.

When installing axle shaft and universal joint assembly exceptional care should be used to prevent damage to the oil seal.

Removing and Overhauling Differential

Inasmuch as the front axle differential assembly is identical with that of the rear axle assembly, refer to the section under "Rear Axle" for the proper procedure to follow in dismantling and assembling differential.

Steering Tie Rod and Bell Crank

These parts being part of the steering mechanism, they are covered in the section under "Steering."

FRONT AXLE TROUBLES AND REMEDIES

SYMPTOMS	PROBABLE REMEDY
Hard Steering	
Lack of lubrication	Lubricate
Tires soft	Inflate to 30 lbs.
Tight steering	Adjust. See "Steering" Section
Low Speed Shimmy or Wheel Fight	
Spring Clips and Shackles loose	Readjust or replace
Front axle shifted	Broken spring center bolt
Insufficient toe-in	Adjust
Improper caster	Reset
Steering System loose or worn	Adjust or overhaul steering gear, front axle or steering parts
Twisted Axle	Straighten or adjust
High Speed Shimmy or Wheel Fight	
Check conditions under "Low Speed Shimmy"	
Tire pressures low or not equal	Inflate to 30 lbs.
Wheels out of balance	Balance—Check for patch
Wheel runout	Straighten
Radial runout of tires	Mount properly
Wheel camber	Same on both wheels
Front springs settled or broken	Repair or replace
Bent steering knuckle arm	Straighten or replace
Shock absorbers not effective	Replace
Steering gear loose on frame	Tighten
Front springs too flexible	Over lubricated
Tramp	
Wheels unbalanced	Check and balance
Wandering	
Improper toe-in	Adjust—Check for bent steering knuckle arm
Broken front spring main leaf	Replace
Axle shifted	Spring center bolt broken
Loose spring shackles or clips	Adjust or replace
Improper caster	Reset
Tire pressure uneven	Inflate to 30 lbs.
Tightness in steering system	Adjust
Loose wheel bearings	Adjust
Front spring settled or broken	Repair or replace

FRONT AXLE TROUBLES AND REMEDIES—Continued

| SYMPTOMS | PROBABLE REMEDY |

Axle Noisy on Pull

Pinion and Ring gear adjusted too tight.............Readjust
Pinion bearings rough...........................Replace

Axle Noisy on Coast

Excessive back lash at ring and pinion gear...........Readjust
End play in pinion shaft........................Readjust
Rough bearing...............................Replace

Axle Noisy on Coast and Pull

Ring and pinion adjusted too tight................Readjust
Pinion set too deep in ring gear..................Readjust
Pinion bearing loose or worn....................Readjust or replace

Back Lash

Axle shaft universal joint worn..................Replace
Axle shaft improperly adjusted ....Readjust
Worn differential pinion washers.................Replace
Worn propeller shaft universal joints..............Replace

Emergency

Where difficulty is experienced with front axle differential making the vehicle inoperative, remove axle driving flanges. This will allow bringing vehicle in under its own power. Be sure front wheel drive shift lever is in the forward (disengaged) position.

FRONT AXLE SPECIFICATIONS

Front Axle

Make..................................Spicer
Drive................................Through springs
Type................................Full Floating
Road Clearance ....8⅞₁₆"

Differential

Drive................................Hypoid
Gear Ratio............................4.88:1
Bearings.............................Timken Roller 2
Adjustment...........................Shims
Gears (Pinion)........................2

Oil Capacity (Pts.)........................See Lubrication Chart, Page 12

Steering Knuckle Thrust Up and Down

Adjusted by shims, should have 25 to 35 inch-pounds pull without oil seal assembly in position.

Steering Knuckle

Bearings Upper and Lower......Timken Roller

Turning Arc........................26°

Tie Rods

Number................................2
Right hand length center to center.......24¼"
Left hand length center to center.......17¹¹⁄₃₂"
Tie rod ends.................Serviced as a unit

Steering Geometry

King Pin inclination....................7½°
Wheel camber........................1½°
Wheel caster.........................3°
Wheel toe-in........................3⁄64"-1⁄32"

Bearings

Cone and Roller.......................24780
Differential side.....................Timken
Cup................................24721
Shims.............003", .005", .010", .030"
Pinion Shaft.........................Timken
  Cone and roller.....Front 31593, Rear 02872
  Cup...........Front 31520, Rear 02820
  Shims............003", .005", .010", .030"

Wheel Hub...:........................Timken
  Cone—Roller.....Inner 18590, Outer 18590
  Cup.........Inner 18520, Outer 18520

Steering Knuckle.....................Timken
  Cone and roller...Upper 11590, Lower 11590
  Cup........Upper 11520, Lower 11520

Steering Bell Crank
  Bearing.........Needle, Torrington B1210

REAR AXLE

The rear axle Fig. 1 is the full-floating type designed so that the axle shafts can be removed without disturbing the wheels. The differential drive is of the hypoid type, having a ratio of 4.88 to 1; 8 tooth drive pinion, 39 tooth drive gear.

The axle shafts are splined at the inner end to fit the splines in the differential side gears. The outer ends of the axle shafts are equipped with integral driving flange which bolts to the rear wheel hubs. The wheels are each supported on two taper roller bearings on the axle housing. The bearing races are pressed into the wheel hub and the adjustment of the bearings made by adjusting nuts on end of housing.

A steel cover is used on the rear of the axle housing to permit inspection and flushing of the differential assembly.

It is necessary to use a hypoid gear lubricant. See Lubrication Chart, Page 12. Various types of hypoid lubricants must not be mixed. If the brand is changed it is best to drain and flush the rear axle housing before installing the new lubricant. The rear axle lubricant level should be checked every 1,000 miles. The lubricant should be drained and axle refilled to the bottom level of the filler hole every 6,000 miles.

Removing Rear Axle from Vehicle

To remove the rear axle first raise the rear end of vehicle with a hoist and support frame ahead of rear springs, then remove wheels and disconnect propeller shaft at rear universal joint by removing U-bolts. Disconnect brake line from hose at frame and remove lock clip. Remove spring clips, then remove spring front bolts, the rear axle can then be removed.

Axle Shaft

To remove the axle shaft, No. 29, Fig. 1, the following procedure should be followed:

1. Remove the six cap screws, No. 35 holding driving flange to wheel hub.

2. Remove axle shaft, two flange screws can be used in the threaded holes in the axle flange.

If rear axle shaft is broken, use a piece of stiff wire and make a loop on one end, slide the wire into axle housing and over broken end of shaft for a sufficient distance that when the wire is pulled out, the loop will bind on shaft and remove it from the differential side gear.

To replace axle shaft, reverse of the above operations are necessary, however, care should be taken when installing shaft that the inner oil seal, No. 28 at differential is not damaged.

Removing and Overhauling Differential

Before disassembling differential, it is advisable to determine through inspection the cause of difficulty or failure of the parts.

Drain lubricant from gear carrier housing and then remove gear carrier cover, No. 6, Fig. 1, flushing out unit thoroughly so that the parts can be carefully inspected.

After the inspection if it is determined that the differential should be completely overhauled, the rear axle assembly should be removed from the vehicle and the following procedure followed:

1. Remove axle shaft as covered in the foregoing paragraph.

2. Remove the four bolts No. 38 which hold the two differential side bearing caps No. 36 in position.

3. Using two pry bars, one on each side of the ring gear parallel with the tube of the axle housing, pry out the differential assembly as shown in Fig. 2.

4. Remove the cap screws No. 3, Fig. 1, holding the bevel drive gear No. 22 on the differential case No. 19.

5. Remove the drive gear from the differential case by tapping it lightly with a lead hammer.

6. The differential shaft No. 13 is held in place by a lock pin No. 25, using a small punch, drive out the lock pin to allow the differential shaft to be removed, Fig. 3.

7. Remove differential pinion gears, No. 15 and 24, Fig. 1, care being taken not to lose the pinion thrust washers, No. 16 and 23.

8. Remove axle shaft gears No. 12 and No. 26 and thrust washers No. 11 and 27.

9. Remove universal joint end yoke assembly No. 59 with puller shown in Fig. 4.

10. With a lead hammer drive on end of pinion shaft which will force the pinion into the gear carrier housing.

11. With bearing race removing tool drive out the front pinion shaft bearing cup No. 65 and oil seal No. 61.

12. To remove the drive pinion rear bearing cone use bearing removing tool or press off in an arbor press, Fig. 5.
    When replacing the cone, select a sleeve the diameter of the cone, so the rollers or cage will not be damaged.

Wash all parts in suitable cleaning fluid, taking care not to lose any of the shims No. 39, Fig. 1 which adjust the pinion shaft bearing running tolerance.

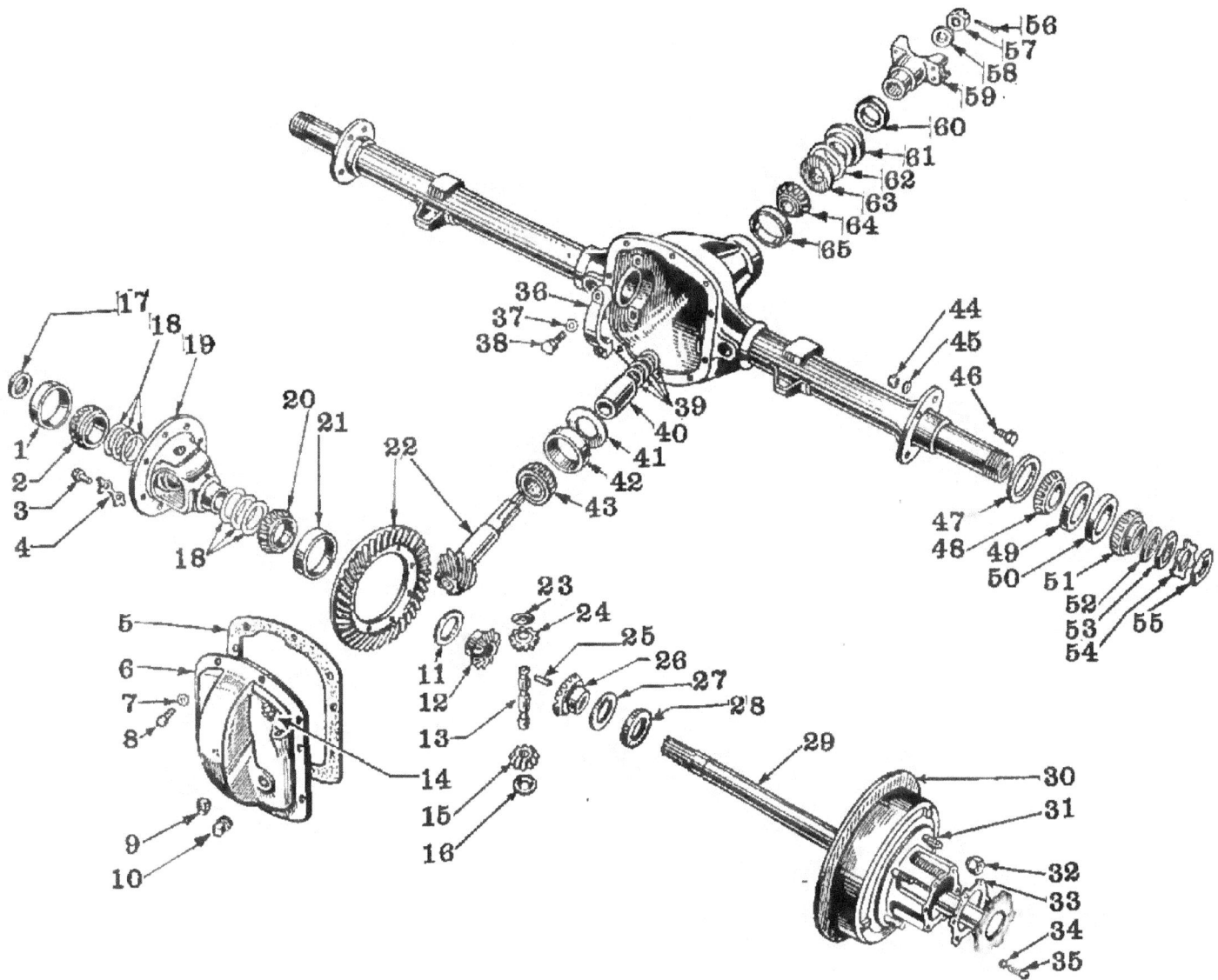

FIG. 1—REAR AXLE ASSEMBLY

No.	Willys Part No.	Ford Part No.	Name	No.	Willys Part No.	Ford Part No.	Name
1	52881	GP-4222	Differential Bearing Cup	32	A-476	GP-1012	Wheel Hub Bolt Nut—R.H. Thread (Ford GP-1013; Willys A-475 L.H. Thread)
2	52880	GP-4221	Differential Bearing Cone and Rollers	33	A-904	GP-4032	Axle Shaft Gasket
3	A-871	355511-S	Hypoid Bevel Drive Gear Screw	34	5010	34807-S	Rear Axle Drive Shaft Screw Lockwasher
4	A-792	GP-4281	Drive Gear Screw Lock Strap	35	A-760	GP-1110	Rear Axle Drive Shaft Screw
5	A-782	GP-4035	Gear Carrier Cover Gasket	36	A-764	GP-4224	Differential Bearing Cap
6	A-781	GP-4016	Gear Carrier Cover	37	636528	34922-S	Differential Bearing Cap Screw Lockwasher
7	52510	34941-S	Gear Cover Screw Lockwasher	38	636527	355699-S	Differential Bearing Cap Screw
8	51523	20346-S2	Gear Cover Screw	39	A-803	GP-4659-A	Pinion Bearing Adjusting Shim (Front)
9	636577	358048-S	Axle Housing Drain Plug	40	A-799	GP-4668	Drive Pinion Bearing Spacer
10	636538	353051-S	Gear Cover Filler Plug	41	A-800	GP-4660-A	Pinion Bearing Adjusting Shim (Rear)
11	A-795	GPW-4228	Differential Bevel Side Gear Thrust Washer	42	52877	86H-4616	Drive Pinion Bearing Cup—(Rear)
12	A-794	GP-4236	Differential Bevel Side Gear	43	52876	86H-4621	Drive Pinion Bearing Cone and Rollers —(Rear)
13	A-798	GP-4211	Differential Bevel Pinion Mate Shaft	44	636575	34083-S2	Brake Disc Screw Nut
14	A-870	GP-4022	Differential Vent Plug	45	5010	34807-S	Brake Disc Screw Lockwasher
15	A-796	GPW-4215	Differential Bevel Pinion Mate	46	A-903	355578-S	Brake Disc Screw
16	A-797	GP-4230	Differential Bevel Pinion Mate Thrust Washer	47	A-854	GP-1177	Hub Oil Seal Assembly
17	A-779	GP-3034	Oil Seal—Carrier End	48	52942	GP-1201	Hub Bearing Cone and Rollers
18	A-784	GP-4229-A	Differential Adjusting Shims	49	52943	GP-1202	Hub Bearing Cup
19	A-793	GP-4206	Differential Case	50	52943	GP-1202	Hub Bearing Cup
20	52880	GP-4221	Differential Bearing Cone and Rollers	51	52942	GP-1201	Hub Bearing Cone and Rollers
21	52881	GP-4222	Differential Bearing Cup	52	A-865	GP-1218	Outer Wheel Bearing Washer
22	A-789	GPW-4209	Hypoid Bevel Drive Gear and Pinion Set	53	A-866	GP-4252	Outer Wheel Bearing Nut
23	A-797	GP-4230	Differential Bevel Pinion Mate Thrust Washer	54	A-867	GP-1124	Outer Wheel Bearing Lockwasher
24	A-796	GP-4215	Differential Bevel Pinion Mate	55	A-866	GP-4252	Outer Wheel Bearing Nut
25	636360	GP-4241	Differential Bevel Pinion Mate Shaft Lock Pin	56	636571	357202-S	Drive Pinion Nut Cotter Pin
26	A-794	GP-4236	Differential Bevel Side Gear	57	636569	356126-S	Drive Pinion Nut
27	A-795	GP-4228	Differential Bevel Side Gear Thrust Washer	58	636570	356504-S	Drive Pinion Nut Washer
28	A-779	GP-3034	Oil Seal Carrier End	59	A-1445	GP-4342	Universal Joint End Yoke Assembly
29	A-901	GPW-4234	Rear Axle Shaft—Right (Ford GP-4235; Willys A-902—Left)	60	636568	GP-4606	Universal Joint End Yoke Dust Shield
				61	639205	GP-4676	Pinion Leather Oil Seal
30	A-472	GP-1111	Brake Drum	62	636565	GP-4661	Pinion Leather Oil Seal Gasket
31	A-474	GP-1107	Wheel Hub Bolt—R.H. Thread (Ford GP-1108; Willys A-473 L.H. Thread)	63	636566	GP-4619	Drive Pinion Oil Slinger
				64	52878	GP-4630	Drive Pinion Bearing Cone and Rollers (Front)
				65	52879	GP-4628	Drive Pinion Bearing Cup (Front)

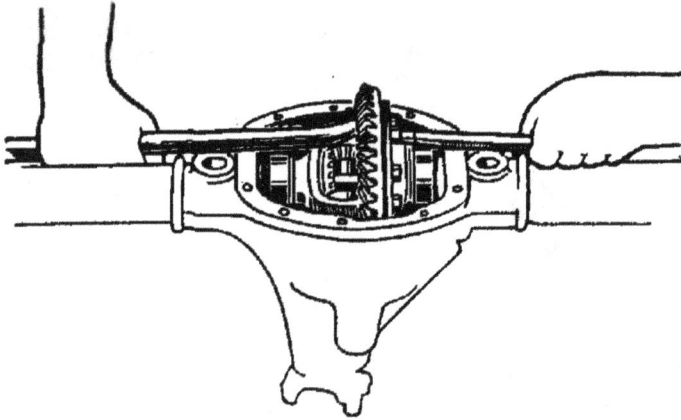

FIG. 2—REMOVING DIFFERENTIAL

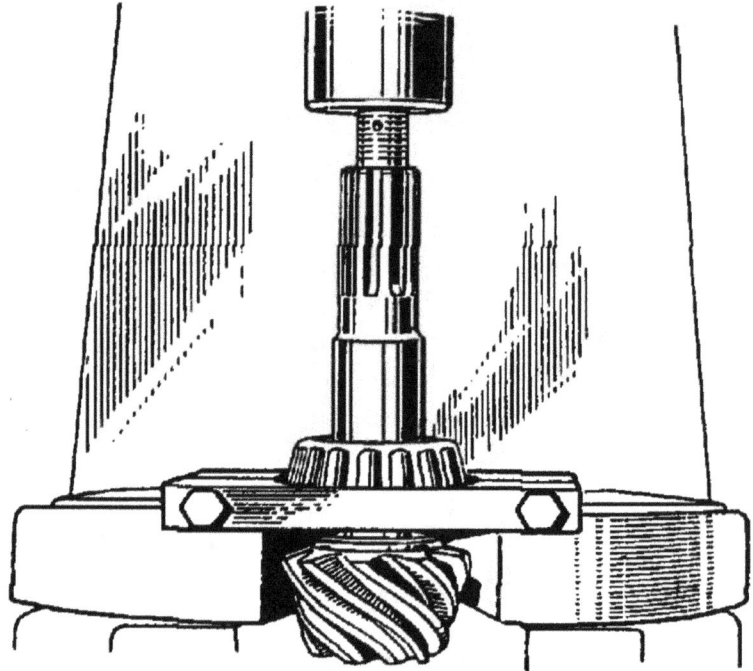

FIG. 5—REMOVING PINION BEARING CONE

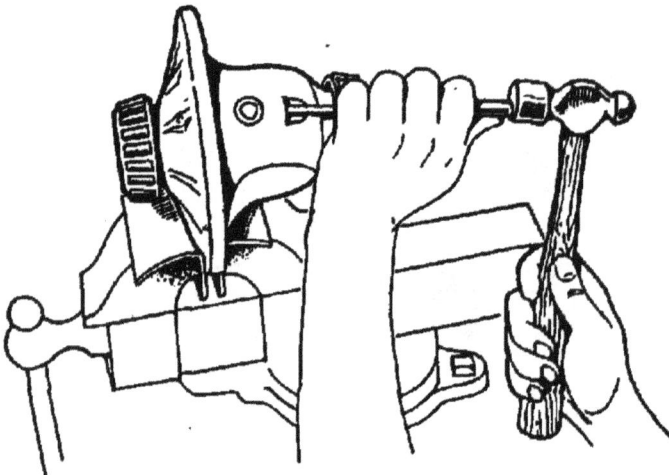

FIG. 3—REMOVING LOCK PIN

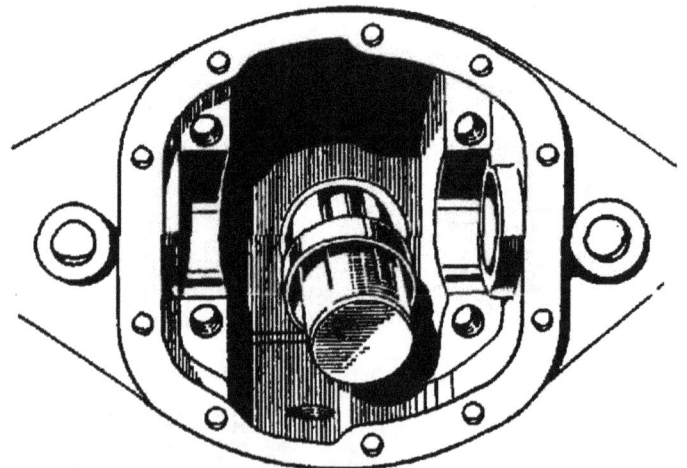

FIG. 6—REMOVING PINION BEARING CUP

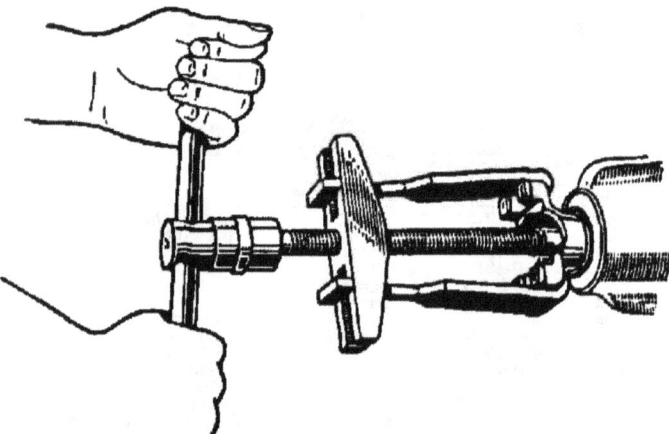

FIG. 4—REMOVING U. J. END YOKE

Adjusting the Drive Pinion

Before attempting to adjust the ring gear or differential parts the drive pinion should be carefully checked and adjusted. The setting of the pinion is accomplished by the use of shims No. 41, Fig. 1 between the rear bearing cup No. 42 and the housing. These shims are available in thickness of .003", .005" and .010".

If the rear bearing cup is to be replaced or if the pinion setting is to be changed, a suitable tool for removing and installing the drive pinion bearing cup in the differential housing should be used, Fig. 6 and 7.

Adjusting Pinion Bearings

The correct pinion bearing adjustment is obtained by shims between the pinion bearing spacer and the front bearing cone, Fig. 8, until a slight drag is obtained when pinion flange is turned by hand.

Install the pinion and the rear bearing in the housing, place the front bearing in position and then install the propeller shaft flange. This operation can be performed very easily by using a block of wood to support the pinion, Fig. 9. Do not install the pinion oil seal until the pinion setting has been checked with the pinion setting gauge.

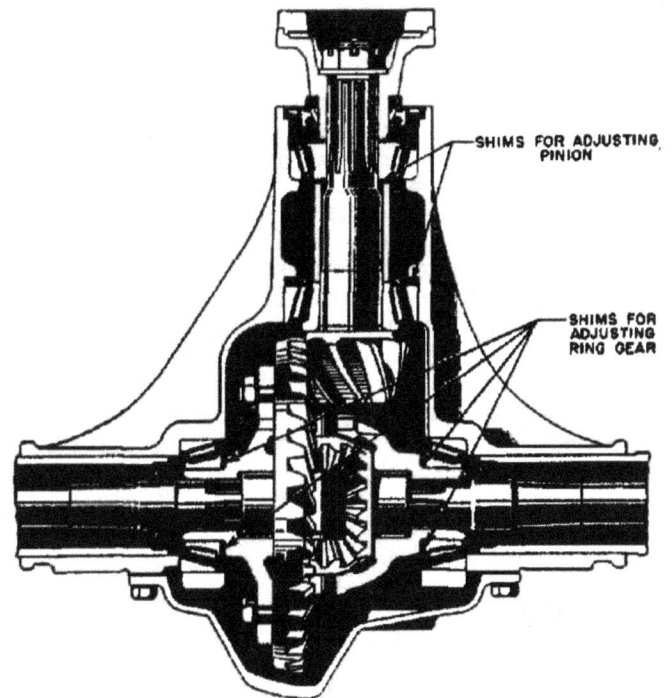

FIG. 8—ADJUSTING SHIMS

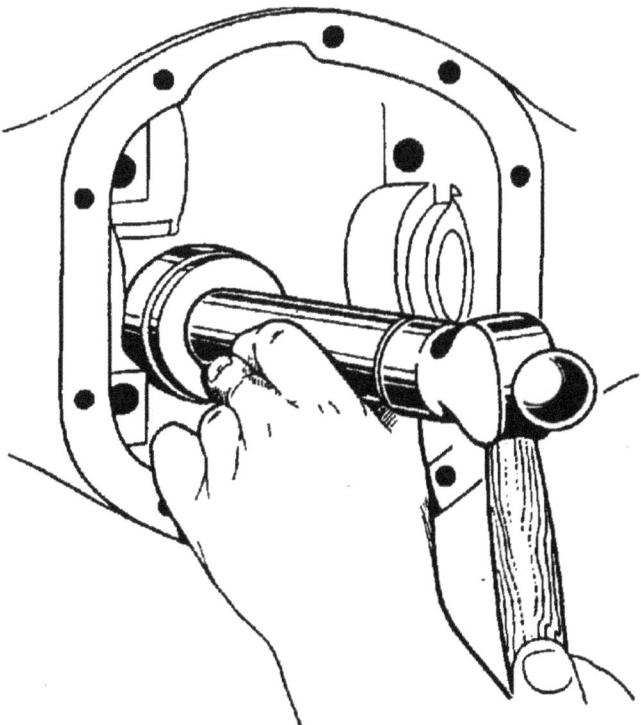

FIG. 7—INSTALLING PINION BEARING CUP

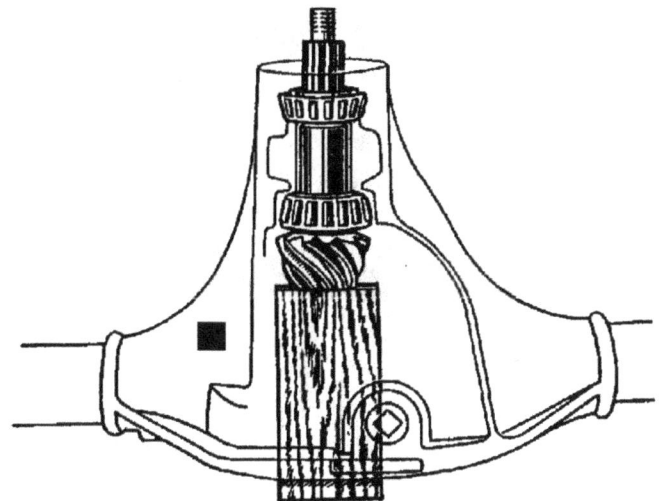

FIG. 9—INSTALLING PINION

Adjusting the Drive Pinion Setting

Proper adjustment of the drive pinion is facilitated by the use of a pinion setting gauge. See Fig. No. 10 and 11. This gauge is fitted with a micrometer for measuring the thickness of the shims required to properly locate the pinion in the differential housing so it will have correct tooth contact with the bevel drive gear.

All axle drive pinions are marked with an electric needle on the back face to show the correct setting. A pinion marked zero will show a reading .719" on the micrometer when properly adjusted. The dimension .719" represents the standard setting from the back face of the pinion to the center line of the differential case bearing. Therefore, a pinion marked +2 is .002" longer than the standard and will show a micrometer reading of .717" when properly adjusted. Likewise a pinion marked —4 is .004" shorter than the standard and will show a micrometer reading of .723" when properly adjusted.

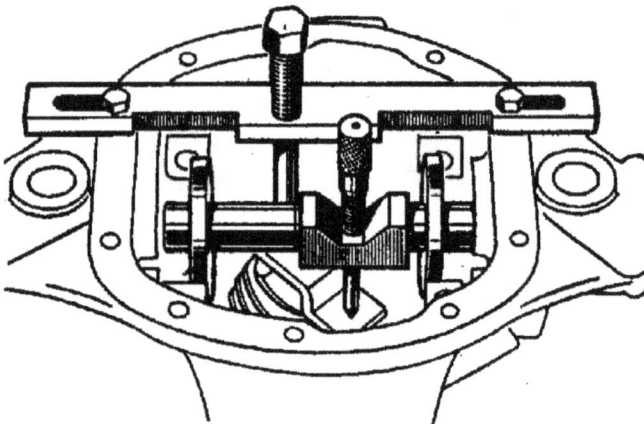

FIG. 10—PINION SETTING GAUGE

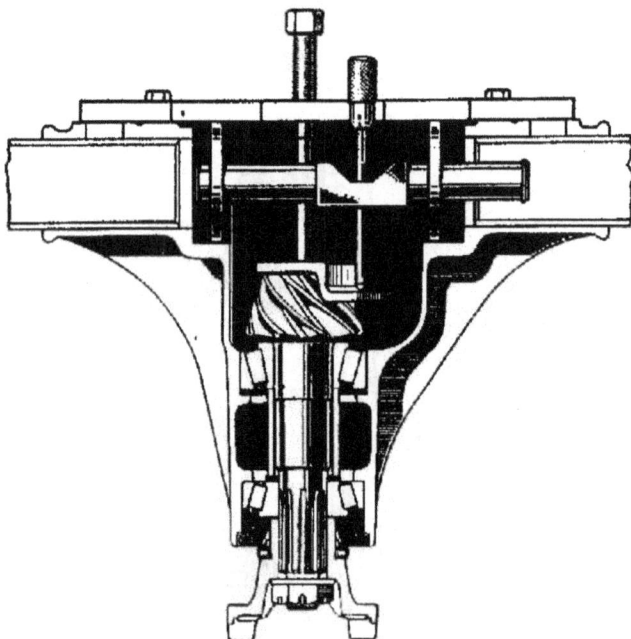

FIG. 11—PINION SETTING GAUGE

Assembling Differential Unit

Carefully examine the surfaces of the differential case and bevel gear to make sure there are no foreign particles or burrs on the two contacting surfaces. Line up the cap screw holes in the bevel gear with those on the differential case and then put it into position on the case by tapping it lightly with a lead hammer. Install the cap screws which hold the bevel gear to the differential case. After the cap screws have been tightened securely, make certain that the cap screw locks are bent around the cap screw heads so there is no possibility of the screws working loose.

The relative assembling position of the internal parts of the differential are shown in Fig. 8. Reassemble the differential pinions, sidegears, thrust washers and shaft in place and install differential shaft lock pin. In order to prevent the lock pin from working out, use a punch to peen over some of the metal of the differential case.

The adjustment of Differential bearings is maintained by the use of Shims between differential case and bearing cones with an .008" pinch fit when assembled in the axle housing.

Remove bearing cones and shims as shown in Fig. 13, reinstall bearing cones without shims, place assembly in Housing with bearing cups and force assembly to one side and check the clearance between bearing cup and case with a feeler gauge as shown in Fig. 12.

After clearance has been determined add .008" this will give thickness of shims required for proper bearing adjustment.

Remove differential bearings and install equal thickness of shims on each side and replace bearings.

Install the differential assembly in the housing. This operation can be facilitated by cocking the bearing cups slightly when the differential is placed in the housing and then tapping them lightly with a lead hammer, see Fig. 14.

FIG. 12—CHECKING DIFFERENTIAL BEARINGS

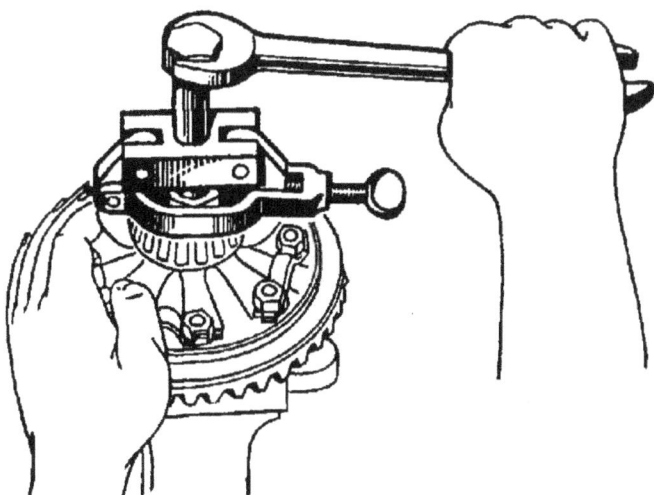

FIG. 13—REMOVING DIFFERENTIAL
BEARING CONE

Total backlash between the bevel gear and pinion should be within .005" to .007". This can be checked by mounting a dial indicator on the rear axle housing with the button of the indicator against one of the gear teeth, Fig. 16. Moving the ring gear by hand will indicate the amount of backlash.

In the event the backlash is not within the limits mentioned above, it will be necessary to change the shims back of the differential case bearings. Changing the position of a .005" shim from one side to the other will change the amount of backlash approximately .0035"

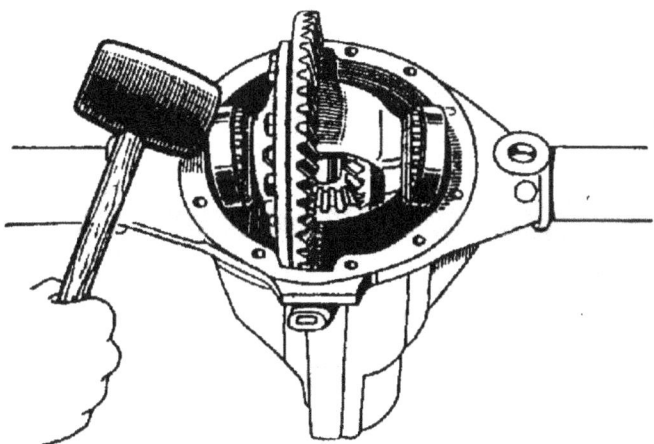

FIG. 14—INSTALLING DIFFERENTIAL

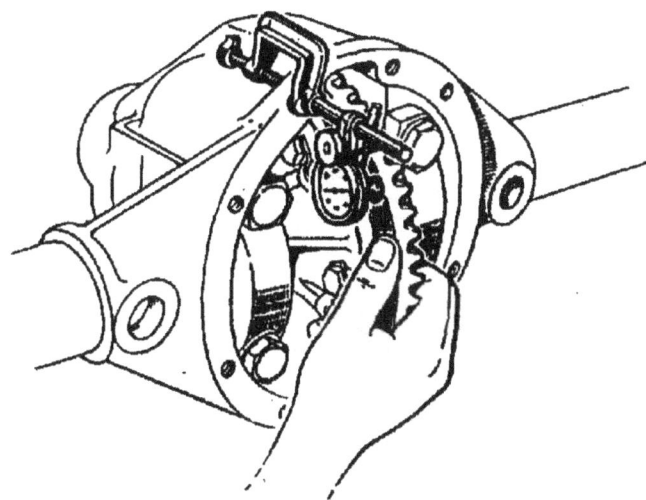

FIG. 15—CHECKING RUNOUT

After the bearing cups are firmly in place in the housing, install the differential bearing caps. It is important that the caps be installed in the same position in which they were originally assembled. Each cap should be installed so numeral corresponds with the numeral on the housing. Torque wrench reading, 38-42 ft. lbs.

After securely tightening the differential bearing caps, check the back face of the ring gear for runout, Fig. 15. Total indicator reading in an excess of .003" indicates a sprung differential case or an improperly installed bevel gear. In either case the assembly must be taken apart and rechecked thoroughly.

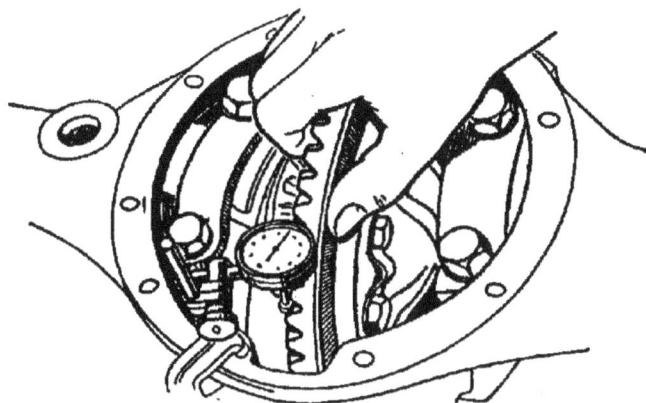

FIG. 16—CHECKING BACK-LASH

In order to assist in determining whether the gears are properly adjusted, paint the bevel gear with red lead or similar substance and turn the bevel gear so the pinion will make an impression on the teeth. Correct procedure to follow in the event of an unsatisfactory tooth contact is shown in Fig. 17.

After the differential has been assembled and adjusted, the pinion shaft oil seal should be installed. Remove universal joint flange and with oil seal replacing tool, Fig. 18 install oil seal. Fig. 19 gives dimensions of oil seal replacing tool. Install universal flange and tighten nut solidly in place, then install cotter pin.

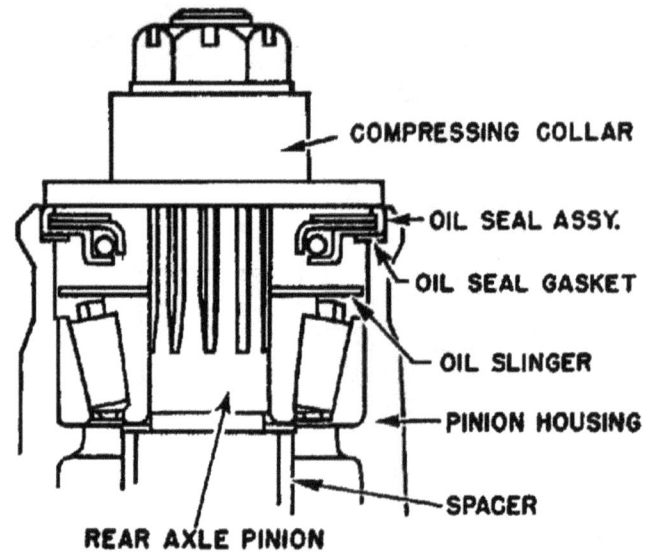

FIG. 18—INSTALLING PINION OIL SEAL

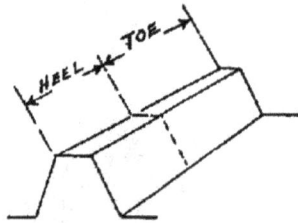

THE HEEL OF GEAR TOOTH IS THE LARGE END, AND THE TOE IS THE SMALL END.

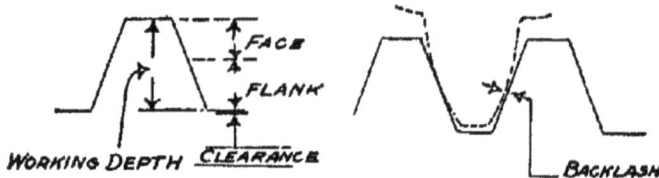

TOO MUCH BACK LASH MOVE GEAR TOWARD PINION

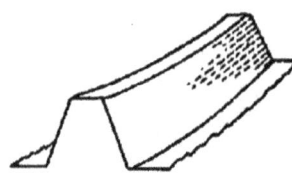

TOO LITTLE BACK LASH MOVE GEAR AWAY FROM PINION.

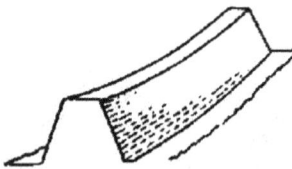

MOVE PINION OUT.

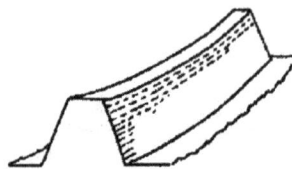

MOVE PINION IN.

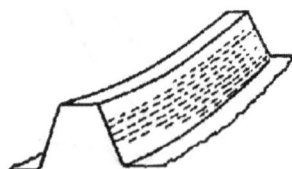

CORRECT SETTING

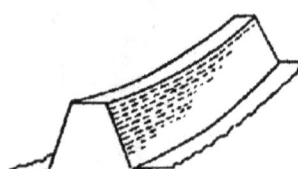

COMPROMISE SETTING

FIG. 17—TOOTH CONTACT

Install axle shafts as instructed under "Axle Shaft" and replace housing cover with new gasket. Fill differential housing with proper amount of hypoid lubricant. See Lubrication Chart, Page 12.

Install axle under vehicle in reverse order of removal, after which bleed the rear brake cylinders to remove any air from the lines, first making certain that there is an ample supply of fluid in the brake master cylinder reservoir. See Section "Brakes" for further instructions.

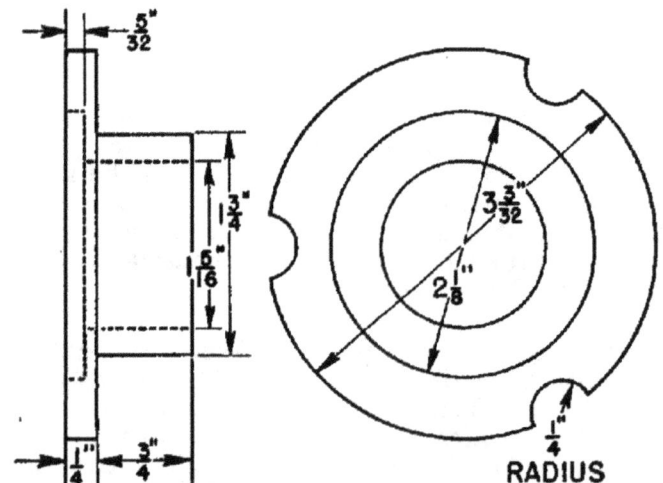

FIG. 19—OIL SEAL COMPRESSING COLLAR

REAR AXLE TROUBLES AND REMEDIES

SYMPTOMS	PROBABLE REMEDY

Axle Noisy on Pull and Coast

Excessive back lash bevel gear and pinion......Adjust
End play pinion shaft.....................Adjust
Worn pinion shaft bearing..................Replace
Pinion set too deep in ring gear.............Adjust
Pinion and bevel gear too tight..............Adjust

Axle Noisy on Pull

Pinion and bevel gear improperly adjusted.....Adjust
Pinion bearings rough ....Replace
Pinion bearings loose......................Adjust

Axle Noisy on Coast

Excessive lash in bevel gear and pinion.......Adjust
End play in pinion shaft....................Adjust
Improper tooth contact....................Adjust
Rough bearings...........................Replace

Backlash

Worn differential pinion gear washers.........Replace
Excessive lash in bevel gear and pinion.......Adjust
Worn universal joints......................Replace

Emergency

Should difficulty be experienced with differential or propeller shaft the vehicle may be driven in by removing the rear axle shafts and propeller shaft.

Place front wheel drive lever in rear (engaged) position. This will allow front wheel drive to propel the vehicle.

REAR AXLE SPECIFICATIONS

Rear Axle

Type.................................Full floating
Make................................Spicer
Drive...............................Thru springs
Road Clearance.......................8⅝"

Differential

Type.................................Hypoid
Ratio...............................4.88:1
Bearings............................Timken Roller
Differential Pinion Gears.............2
Oil capacity.........................See Lubrication Chart, Page 12
Adjustment..........................Shims .003", .005", .010", .030"

Pinion Shaft

Bearings............................Two Timken Roller
Adjustment..........................Shims .003", .005", .010"

Bevel and Pinion Gear

Back Lash...........................005"—.007"
Adjustment..........................Shims .003", .005", .010", .030"

Bearings

Make—Differential Side................		Timken
Cone and roller......................		24780
Cup................................		24721
Make—Pinion Shaft...................		Timken
Cone and roller......................	Front 02872	Rear 31593
Cup................................	Front 02820	Rear 31520
Shims..............................	.003", .005", .010", .030"	
Make—Wheel Hub....................		Timken
Cone and Roller......................	Inner 18590	Outer 18590
Cup................................	Inner 18520	Outer 18520

BRAKES

The brake system is comparatively simple, Fig. 1. The foot or service brakes are of the internal expanding type hydraulically actuated in all 4 wheels. The hand brake is mechanically operated through a cable and conduit to an external type brake mounted at the rear of the transfer case on the propeller shaft. The foot brakes are of the Bendix, two shoe, double anchor type and have nickle-chromium alloy iron drums.

In order to thoroughly understand the operation of the hydraulic brake system, it is necessary to have a good knowledge of the various parts and their functions, and to know what takes place throughout the system during the application and the release of the brakes.

The piston in the master cylinder, Fig. 4 receives mechanical pressure from the brake pedal and exerts pressure on the fluid in the lines, building up the hydraulic pressure which moves the wheel cylinder pistons. The primary cup is held against the piston by the piston return spring which also holds the check valve against the seat. The spring maintains a slight fluid pressure in the line and in the wheel cylinders to prevent the possible entrance of air into the system. The secondary cup which is secured to the opposite end of the piston, prevents the leakage of fluid into the rubber boot. The holes in the piston head are for the purpose of allowing the fluid to flow from the angular space around the piston into the space between the primary cup and the check valve, keeping sufficient fluid in the line at all times. The holes in the check valve case allow the fluid to flow through the case, around the lips of the rubber valve cup and out into the line during the brake application. When the brakes are released the valve is forced off the seat permitting the fluid to return to the master cylinder. The piston assembly is held in the opposite end of the housing by means of a snap ring. The rubber boot that fits around the push rod and over the end of the housing prevents dirt or any foreign matter from entering the master cylinder.

The wheel cylinder is a double piston cylinder, the purpose of the two pistons being to distribute the pressure evenly to each of the two brake shoes. Rubber piston cups maintain pressure on the pistons to prevent the leakage of fluid. The rubber boots over the end of the cylinder prevent dust and dirt or foreign material from entering the cylinder.

When pressure is applied to the brake pedal, the master cylinder forces fluid through the lines and into the wheel cylinders. The pressure forces the pistons in the wheel cylinder outward, expanding the brake shoes against the drum. As the pedal is further depressed higher pressure is built up within the hydraulic system, causing the brake shoes to exert a greater force against the brake drums.

As the brake pedal is released, the hydraulic pressure is released and the brake shoe retracting spring draws the shoes together, pushing the wheel cylinder pistons inward and forcing the fluid out of the cylinder back into the line towards the master cylinder. The piston return spring in the master cylinder returns the piston to the piston stop faster than the brake fluid is forced back into the line, which creates a slight vacuum in that part of the cylinder ahead of the piston. The vacuum causes a small amount of fluid to flow through the holes in the piston head, past the lip of the primary cup and into the forward part of the cylinders. This action keeps the cylinder filled with fluid at all times, ready for the next brake application. As fluid is drawn from the space behind the piston head it is replenished from the reservoir through the intake port. When the piston is in the fully released position the primary cup clears the bypass port, allowing the excess fluid to flow from the cylinder into the reservoir as the brake shoe retracting springs force the fluid back into the master cylinder.

Brake Pedal Adjustment

There should always be at least ½" free pedal travel before the push rod engages the piston.

This adjustment is accomplished by the shortening or lengthening of the brake master cylinder eye bolt, No. 59, Fig. 1. This is done so the primary cup will clear the port No. 15, Fig. 4, when the piston is in the off position, otherwise the compensating action of the master cylinder for expansion and contraction of the fluid in the system, due to temperature changes, will be destroyed and cause the brakes to drag.

Brake Shoe Adjustment—Minor

When the brake lining becomes worn, as indicated by foot pedal going almost to the floor board, necessary adjustment can readily be made as described in the following paragraph; first making certain that there is ½" free brake pedal travel.

Jack up the wheels to clear the floor. Adjustment is made by rotating the eccentric No. 5, Fig. 1. With a wrench loosen lock nut No. 6 for forward brake shoe, hold lock nut and with another wrench turn eccentric towards the front of the car until brake shoe strikes drum, then turning wheel with one hand release eccentric until wheel turns free, holding eccentric tighten lock nut. To adjust reverse shoe, repeat this operation only turn the eccentric towards the back of the car. Do this on all brakes. Check fluid in master cylinder.

FIG. 1—BRAKE SYSTEM

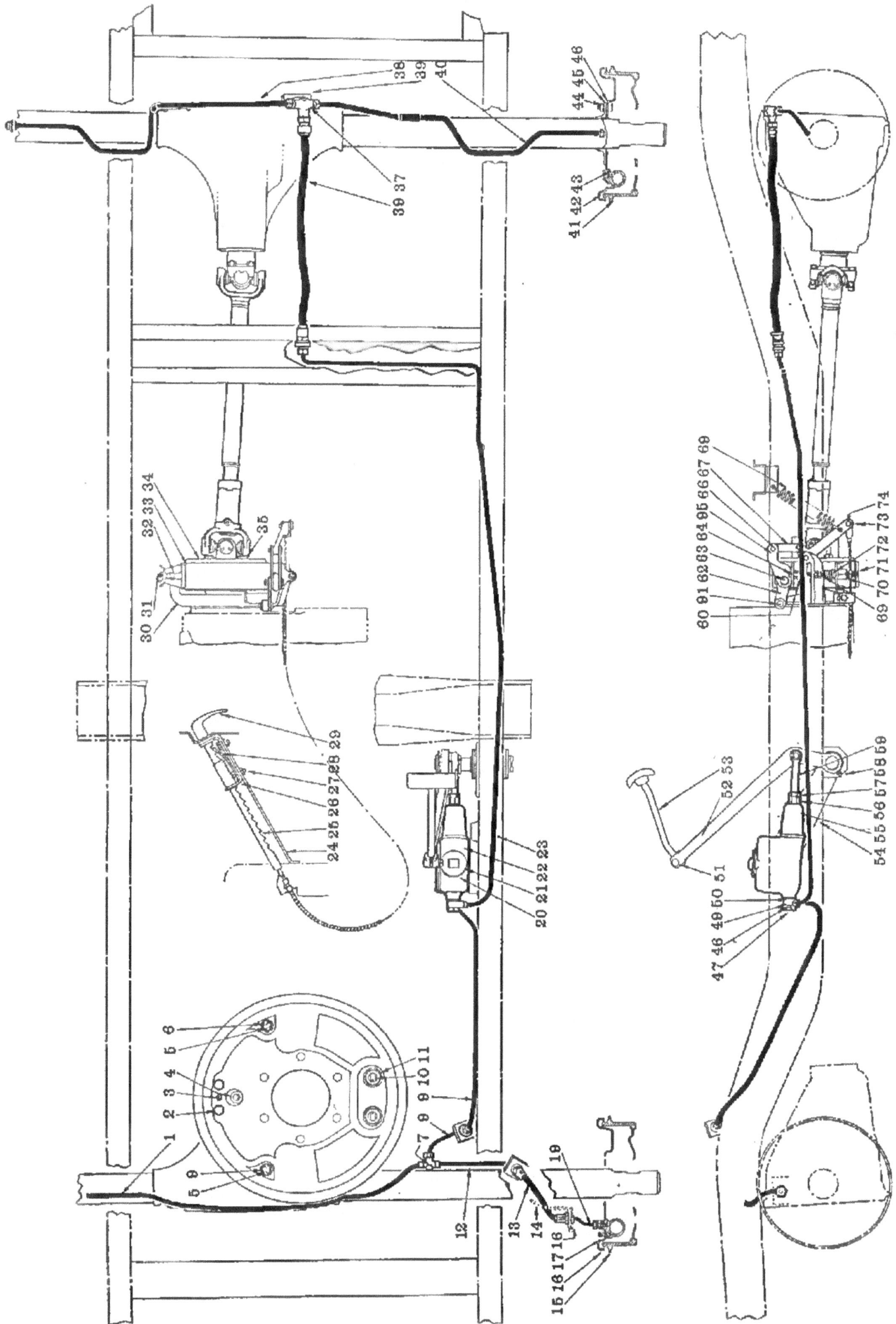

FIG. 1—BRAKE SYSTEM

No.	Willys Part No.	Ford Part No.	Name
1	A-1376	GPW-2266	Brake Tube Assembly (Tee to Front Brake Hose—Right)
2	51738	20300-S7	Hex. Head Screw (Wheel Cylinder to Backing Plate)
3	637540	GP-2208	Wheel Brake Cylinder Bleeder Screw
4	A-1502		Front Wheel Brake Cylinder
5	A-754	GP-2038	Brake Shoe Eccentric
6	A-755	33800-S7-8	Hex. Nut (Eccentric)
7	637432	GP-2074	Axle Tee
8	A-1373	GPW-2078	Brake Hose—Front (Axle to Frame)
9	A-1377	GPW-2264	Brake Tube Assembly (Master Cylinder to Front Hose)
10	637899	91A-2027	Brake Shoe Anchor Pin
11	637924	33846-S7-8	Hex. Nut (Anchor Pin)
12	A-1501	GPW-2203	Brake Tube Assembly (Tee to Front Brake Hose—Left)
13	A-1460	GPW-2079	Brake Hose (Front Axle)
14	A-1457	GPW-2096	Front Wheel Brake Hose Guard
15	A-472	GPW-1125	Front Brake Drum
16	A-450	GP-2013	Front Brake Backing Plate
17	637540	GP-2208	Wheel Brake Cylinder Bleeder Screw
18	637427	78-2314-A	Spring Lock Clip (Brake Hose to Bracket)
19	A-1488	GPW-2298	Brake Tube Assembly (Wheel Cylinder to Axle Hose—Left)
20	637612	GP-2167	Master Cylinder Filler Cap Gasket
21	637608	GP-2162	Master Cylinder Filler Cap Assembly
22	637582	GP-2155	Master Cylinder and Supply Tank
23	A-5224		Brake Tube Assembly (Master Cylinder to Rear Hose)
24	A-2892	GPW-2852	Hand Brake Ratchet Tube Bracket Support
25	A-1242	GPW-2780	Hand Brake Handle Tube and Cable Assembly
26	639010	GPW-2848	Hand Brake Ratchet Tube Bracket Assembly
27	51396	24347-S	Hex. Head Screw (Bracket to Support)
28	635681	GPW-2793	Hand Brake Ratchet Tube Spring
29	639244	GPW-2782	Hand Brake Handle
30	A-1507	GPW-7769	Transfer Case Output Shaft Bearing Cap—Rear
31	A-1020	O1T-2616	Hex. Head Screw (Anchor Clip)
32	A-1021	O1T-2640	Transmission Brake Band Anchor Clip Screw Spring
33	A-1009	GP-2648	Transmission Brake Band and Lining Assembly
34	A-1002	GP-2614	Brake Drum (Transmission)
35	636575	33786-S	Hex. Nut (Brake Drum to Flange and Propeller Shaft)
36	637424	GP-2078	Brake Hose—Rear (Axle to Frame)
37	637432	GP-2074	Axle Tee
38	A-5226		Brake Tube Assembly (Tee to Rear Brake—Right)
39	A-5227		Rear Axle Brake Tube Tee Bracket
40	A-5225		Brake Tube Assembly (Tee to Rear Brake—Left)
41	A-472	GP-1111	Rear Brake Drum
42	A-450	GP-2013	Rear Brake Backing Plate Assembly
43	A-6111		Rear Wheel Brake Cylinder
44	A-903	355578-S	Brake Backing Plate Screw
45	636575	33786-S2	Brake Backing Plate Screw Nut
46	5010	34807-S7-8	Brake Backing Plate Screw Lockwasher
47	637505	GP-2077	Brake Master Cylinder Outlet Fitting Bolt
48	637606	91A-2151	Outlet Fitting Gasket—Large
49	A-557	GP-2076	Outlet Fitting
50	637604	91A-2152	Outlet Fitting Gasket—Small
51	6157	24426-S2	Clamp Screw (Pedal Shank to Pedal)
52	A-1388	GPW-2452	Brake Pedal Assembly
53	A-1369	GPW-2454	Brake Pedal Pad Assembly
54	A-1354	GPW-2138	Master Cylinder Tie Bar
55	637602	GP-2180	Brake Master Cylinder Boot
56	637599	GP-2143-A1-2	Brake Master Cylinder Push Rod Assembly
57	5939	33802-S	Brake Master Cylinder Eye Bolt Lock Nut
58	392909	353027-S7-S	Brake Pedal Hydraulic Fitting
59	A-163	GPW-2462	Brake Master Cylinder Eye Bolt
60	A-1017	O1T-2634	Transmission Brake Releasing Spring (Hand)
61	A-1006	73889-S7	Transmission Brake Support Quadrant Pin (Hand)
62	A-1005	GPW-2630	Transmission Brake Support Quadrant (Hand)
63	A-1004	73928-S7-8	Transmission Brake Cam Pin (Hand)
64	A-1019	31218-S7	Transmission Brake Band Bracket Cap Screw (Hand)
65	A-1003	GPW-2632	Transmission Brake Cam (Hand)
66	311003	73904-S7	Clevis Pin (Link to Cam)
67	A-1228	GPW-2559	Hand Brake Relay Crank Link
68	A-5335	GPW-2635	Hand Brake Retracting Spring
69	5790	33795-S	Brake Band Cap Screw Nut
70	A-1018	O1T-2805	Transmission Brake Band Adjusting Nut (Hand)
71	A-1016	O1T-2842	Transmission Brake Adjusting Bolt (Hand)
72	A-1017	O1T-2634	Transmission Brake Releasing Spring (Hand)
73	392468	357553-S18	Clevis Pin (Cable to Relay Crank)
74	A-1226	GPW-2656	Hand Brake Relay Crank Assembly

Brake Shoe Adjustment—Major

In the event the minor adjustment does not give adequate brakes or when it is necessary to reline the brakes it will be necessary to reset anchor pins, No. 10. The brake adjustments should be made as follows:

With the shoe and lining assemblies installed and the adjusting fixture or brake drum in place, loosen the anchor pin lock nuts No. 11 on the rear of the backing plate. Adjustment is made by turning the eccentric anchor pins towards each other and down until the shoes are set to the proper clearance, as determined by feeler gauges. The recommended shoe setting is .005" clearance at the heel (lower end), and .008" at the toe (upper end) of brake shoe lining. A slot is provided in the brake drum for checking these clearances.

Relining Brake Shoes

When necessary to reline the brakes, the car should be raised so that all four wheels are free from the floor.

Remove the wheels and then the hubs and drums which will then give access to the brake shoes.

Install wheel cylinder clamps or keepers to retain the wheel cylinder pistons in place and prevent leakage of brake fluid while replacing the shoes. Turn all eccentrics to the lowest side of the cam, and then remove the brake shoe contracting spring. No. 1, Fig. 2.

Remove anchor pin nuts, lock washers, and anchor pins from backing plate.

Remove rivets holding lining to shoes and install new linings through the use of a brake lining clamp.

Inspect the oil seals in the wheel hubs and if found that grease has been leaking, it is advisable to install new oil seals.

Install brake shoes to the brake backing plate, the shoe with the longest lining is the forward shoe on all four wheels. Install anchor pin No. 12, pin plate No. 2 and pin cam No. 13; then install anchor pins so the punch mark on the ends are facing each other. Install lock washer and nut. Install brake shoe return spring. Remove brake cylinder clamp or keeper.

Install the hubs and drums, then make the major adjustment of the brakes.

If it is found while the wheels are removed that there is brake fluid leakage at any of the wheel cylinders, it will be necessary to recondition that wheel cylinder, and bleed the brake lines. This subject is covered under the heading "Reconditioning Wheel Cylinders" and "Master Cylinder".

NOTE: Whenever the brake lining is renewed in one front or rear wheel be sure to perform the same operations in the opposite front or rear wheel, using the same brake lining as to color and part number, otherwise unequal brake action will result.

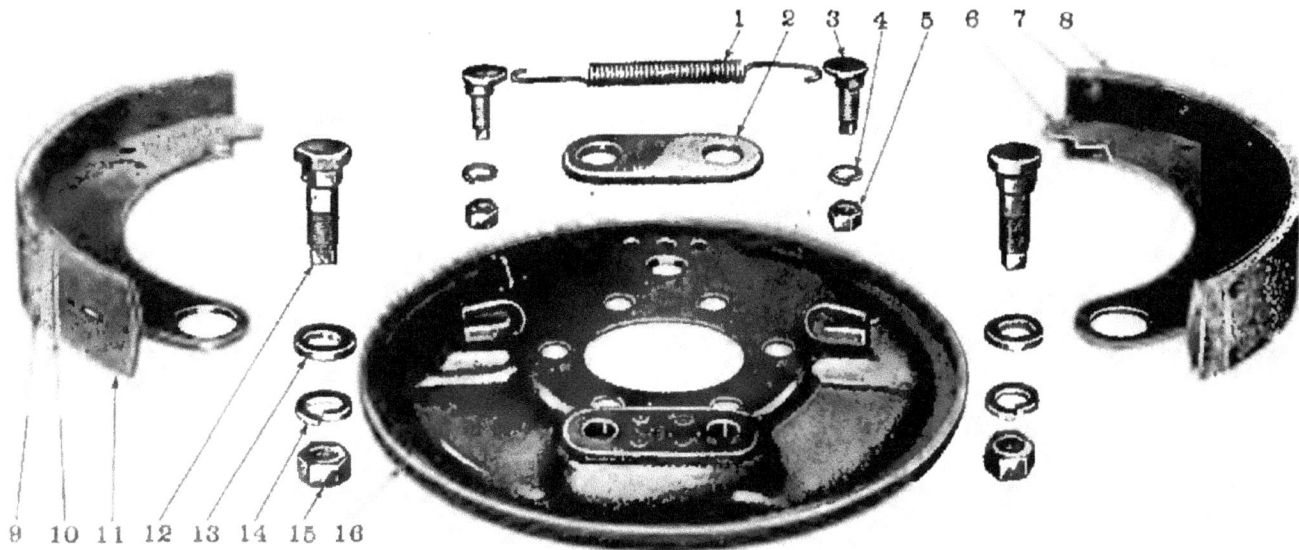

FIG. 2—BRAKE

No.	Willys Part No.	Ford Part No.	Name	No.	Willys Part No.	Ford Part No.	Name
1	637905	GP-2035	Brake Shoe Return Spring	9	374586	351915-S	Brake Lining Tubular Brass Rivet
2	637901	91A-2030	Brake Shoe Anchor Pin Plate	10	116552	GP-2022	Brake Shoe Lining—Reverse
3	A-754	GP-2038	Brake Shoe Eccentric	11	116550	GP-2019	Brake Shoe and Lining Assembly—Reverse
4	5010	34807-S7-8	Brake Shoe Eccentric Lockwasher				
5	A-755	33800-S7	Brake Shoe Eccentric Nut	12	637899	91A-2027	Brake Shoe Anchor Pin
6	116549	GP-2018	Brake Shoe and Lining Assembly—Forward	13	637900	GP-2028	Brake Shoe Anchor Pin Cam
				14	637923	351466-S24	Brake Shoe Anchor Pin Lockwasher
7	374586	351915-S	Brake Lining Tubular Brass Rivet	15	637924	33846-S7-8	Brake Shoe Anchor Pin Nut
8	116551	GP-2021	Brake Shoe Lining—Forward	16	A-450	GP-2013	Brake Backing Plate Assembly

Bleeding Brakes

The hydraulic brake system must be bled whenever a fluid line has been disconnected or air gets into the system. A leak in the system may sometimes be evidenced through the presence of a spongy brake pedal. Air trapped in the system is compressible and does not permit pressure applied to the brake pedal to be transmitted solidly through to the brakes. The system must be absolutely free from air at all times. When bleeding the brakes it is advisable that the longest fluid line from the master cylinder be bled first. The proper sequence of bleeding is right rear; right front; left rear; left front. During the bleeding operation the master cylinder must be kept at least ¾ full of hydraulic brake fluid.

To bleed the brakes first carefully clean all dirt from around the master cylinder filler plug. Remove filler plug and fill master cylinder to the lower edge of filler neck. Clean off all bleeder connections at all four wheel cylinders. Attach bleeder hose and fixture to right rear wheel cylinder bleeder screw and place end of tube in a glass jar, end submerged in fluid. Open the bleeder valve ½ to ¾ of a turn. See Fig. 3.

Depress the foot pedal by hand, allowing it to return very slowly. Continue this pumping action to force the fluid through the line and out the bleeder hose which carries with it any air in the system.

When bubbles cease to appear at the end of the bleeder hose, tighten the bleeder valve and remove the hose.

After the bleeding operation has been completed at all four wheels, fill the master cylinder reservoir and replace the filler plug.

It is not advisable to re-use the fluid which has been removed from the lines through the bleeding process.

Master Cylinder

When necessary to remove the master cylinder No. 22, Fig. 1 for reconditioning, the following procedure should be followed:

1. Raise front end of car with jack because the removal operation must all be performed from the under side of the vehicle.
2. Remove stop light switch wires.
3. Remove fitting bolt and switch No. 47.
4. Remove front bolt holding master cylinder to frame which is installed from the outside of the frame and screws into master cylinder body.
5. Remove master cylinder tie bar cap screw which is the front inside screw on master cylinder.
6. Remove rear master cylinder bolt which has nut on inside of frame bracket.
7. Remove master cylinder boot No. 55.
8. Remove master cylinder.

The installation of master cylinder to frame is the reverse of the above operations.

After the master cylinder has been removed it should be dismantled and thoroughly washed in alcohol. (Never wash any part of the hydraulic braking system with gasoline (petrol) or kerosene.)

Bleed lines as instructed under the heading, "Bleeding Brakes". See Fig. 3.

Recheck entire system to make sure that there are no leaks and if necessary make brake adjustments in order to have adequate brakes.

Filling Master Cylinder

The Master Cylinder reservoir should be checked each 1000 miles when vehicle is lubricated, be sure that there is a sufficient supply of brake fluid. The master cylinder can be reached by removing the five screws in the inspection cover on the toeboard below the steering post. After removing the cover thoroughly clean any dirt away from the filter cap on the master cylinder to prevent it from entering the brake system where it might cause a scored cylinder or possible failure of the brakes.

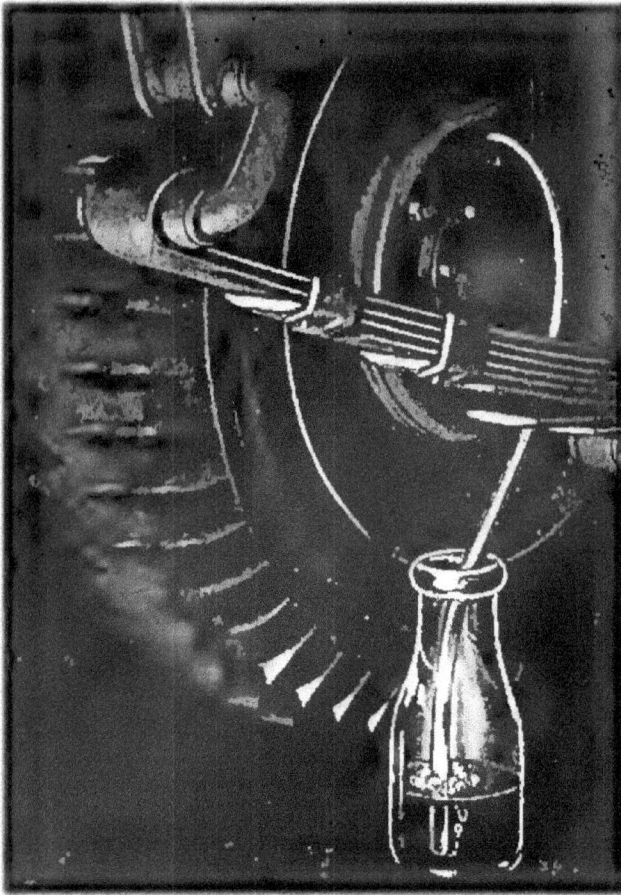

FIG. 3—BLEEDING BRAKE

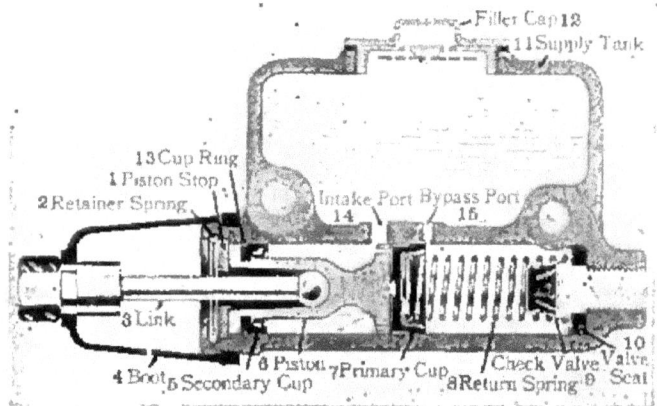

FIG. 4—MASTER CYLINDER

After the parts have all been thoroughly cleaned with alcohol, make careful inspection, replacing those parts which show signs of being deteriorated. Inspect cylinder bore and if found to be rough it should be honed out. The clearance between the piston and the cylinder bore should be .001" to .005". During the honing operation use hydraulic brake fluid on the hone, in order to obtain a polished surface in the cylinder bore. Wash out cylinder with alcohol and with a wire passed through the ports No. 14 and No. 15 that open from the supply reservoir into the cylinder bore, make sure that these passages are free and clear of any foreign matter. It is our recommendation that a new piston, primary cup, valve and valve seat be installed when rebuilding the master cylinder. See Fig. 4.

Install valve seat No. 10 in end of cylinder with flat surface toward valve. Install valve assembly No. 9. Install piston return spring No. 8. Install primary cup No. 7. Face of cup goes towards piston. Install piston No. 6, stop plate No. 1 and lock wire No. 2. Install fitting connection with new copper gaskets. Fill reservoir half full of brake fluid and operate the piston with piston rod until fluid is ejected at fitting. Install master cylinder to frame and make necessary connections and adjust pedal clearance to ½" free play.

Wheel Cylinders

To remove a hydraulic brake wheel cylinder Fig. 5, jack up the vehicle and remove the hub and drum. Disconnect brake line at fitting on brake backing plate. Remove brake shoe retracting spring which allows the brake shoes at the toe to fall clear of the brake cylinder. Remove two cap screws holding wheel cylinder to backing plate.

Remove rubber dust covers on end of cylinders, then pistons, piston cups and spring.

Wash the parts in clean alcohol. Examine the cylinder bore for roughness or scoring. Check fit of pistons to cylinder bore by using .002" feeler gauge.

In reassembling cylinder dip spring, pistons and piston cups in brake fluid. Install spring in center of wheel cylinder. Install piston cups with the cupped surface towards the spring so that the flat surface will be against piston. Install pistons and dust covers. Install wheel cylinder to backing plate, connect up brake line and install brake shoe retracting spring. Replace wheel hub and drum, then bleed the lines as instructed under "Bleeding Brakes."

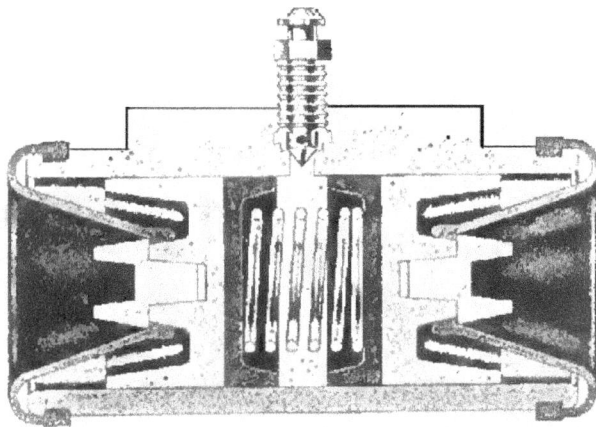

FIG. 5—WHEEL CYLINDER

Brake Hose—Front

To remove brake hose at the wheels the following procedure should be followed to prevent damage to hose and fitting. See Fig. 6.

1. Remove brake line connections at each end.
2. Slip brake hose spring lock clip off ends of hose fitting and remove brake hose from brackets.

To remove front brake hose, frame to axle the following procedure should be followed:

1. Remove brake line connection on frame bracket, top connection.
2. Remove brake hose spring lock clip from brake hose fitting at bracket.
3. Remove brake hose from bracket.
4. With open end wrench unscrew brake hose from tee connection on axle and remove.

FIG. 6—BRAKE HOSE

Brake Hose—Rear

To remove the rear brake hose, the following procedure should be followed:

1. Remove brake line from hose connection at frame.

2. Slip brake hose spring lock clip off of brake hose fitting.
3. Remove brake hose from frame.
4. With open end wrench unscrew brake hose from fitting on axle housing.

Whenever a brake line has been disconnected, it will be necessary to bleed the brakes. The bleeding of the brakes should be done in accordance with instructions given under "Bleeding Brakes".

Transmission Hand Brake

The hand brake is applied to the rear propeller shaft at transfer case, see Fig. 1. The operation of the brake is positive through a cable connection.

To adjust the hand brake Fig. 7, the following operations should be performed.

Have hand brake lever on dash in the released position. Adjust anchor screw No. 1, so that there will be .005"—.010" clearance between the band and the drum. Tighten nut No. 2 until band is brought tight against the drum. Adjust bolt No. 3 so that the head just rests on upper half of the band. Back off two turns on adjusting nut No. 2.

The length of the cable from the hand grip to the brake levers is of a predetermined length and cannot be changed. At the regular lubrication periods of 1,000 miles it is advisable to put a few drops of oil in the upper end of conduit tube at cable to keep it free to slide within the conduit.

This brake is designed for holding the car while parked.

To reline brake band, remove from bracket and adjusting linkage. Cut off rivets and remove lining, care being taken not to distort the band. Hold end of new lining flush with end of band, make lining hug band inside and cut off about 5/16"-1/2" long. Bring lining ends even with ends of band and install end rivets only. Then remove center bulge in lining with hammer, so lining will hug band tightly, then install balance of rivets and form band to drum, making regular adjustments.

FIG. 7—TRANSMISSION BRAKE

BRAKE TROUBLES AND REMEDIES

SYMPTOMS	PROBABLE REMEDY

Brakes Drag

Brake shoes improperly adjusted...............	Readjust
Piston cups—Enlarged.........................	Flush all lines with alcohol—Install new cups in Wheel and Master cylinders
Mineral oil or improper brake fluid in system...	
Improper pedal adjustment....................	Adjust master cylinder rod
Clogged master cylinder compensating port......	Clean master cylinder

One Brake Drags

Brake shoe adjustment incorrect...............	Adjust
Brake hose clogged...........................	Replace
Retracting spring broken......................	Replace
Wheel cylinder piston or cups defective........	Replace
Loose or damaged wheel bearings.............	Adjust or replace

Brake Grabs—Car Pulls to One Side

Brake anchor pin adjustment incorrect..........	Adjust
Oil or brake fluid on lining...................	Replace lining
Dirt between lining and drum.................	Clean with wire brush
Drum scored or rough........................	Turn drum and replace lining
Loose wheel bearings.........................	Adjust
Axle spring clips loose.......................	Tighten
Brake backing plate loose.....................	Tighten
Brake lining.................................	Different kinds on opposite wheels
Brake shoes reversed.........................	Primary and secondary shoes reversed in one wheel
Tires under-inflated..........................	Inflate to 30 lbs. pressure
Tires worn unequally.........................	Replace or change around to opposite wheels

Excessive Pedal Travel

Normal lining wear...........................	Adjust
Lining worn out..............................	Replace
Leak in brake line............................	Locate and repair
Scored brake drums..........................	Replace or regrind
Incorrect brake lining........................	Replace
Air in hydraulic system.......................	Fill master cylinder and Bleed lines

Spongy Brake Pedal

Air in lines..................................	Bleed lines
Brake shoe adjustment incorrect..............	Adjust

Excessive Pedal Pressure

Oil or brake fluid on lining...................	Replace lining
Shoes improperly adjusted....................	Major adjustment
Warped brake shoes..........................	Replace
Distorted brake drums........................	Replace or regrind

Squeaky Brakes

Brake shoes warped or drums distorted........	Replace
Lining loose.................................	Replace
Dirt imbedded in lining.......................	Clean with wire brush or replace
Improper adjustment.........................	Adjust

BRAKE SPECIFICATIONS

Service Brakes:

Type.....................4 Wheel Hydraulic
Size..........................9" x 1¾"
Fluid Capacity Pts. See Lubrication Chart, Pg. 12

Master Cylinder:

Size............................1"
Mounted...............L.H. Frame Side Rail

Wheel Cylinder:

Size.................Front 1"   Rear ¾"

Brake Shoes....................Bendix
Size.........................9" x 1¾"
Lining area..................117.8 Sq. in.
Length Lining-Forward shoe..........10⅛₂"
Length Lining-Reverse shoe........6³⁹⁄₆₄"
Width........................1¾"
Thickness......................⁵⁄₁₆"

Hand Brake

Type...........................Mechanical
Size..6"
Lining..................................Woven
Length............................18⁹⁄₁₆"
Width....................................2"

Brake Return Springs:

Brake Pedal
    Free Length...........................5⅞"
    Load when extended to 7⁹⁄₁₆".......23 lbs.

Brake Shoe Return Spring
    Free Length.........................5¹³⁄₁₆"
    Load when extended to 6³⁄₁₆".......40 lbs.

Wheel Cylinder Spring
    Length............................1⁷⁄₁₆"
    Load when compressed........1 to 1¼ lbs.

WHEELS—WHEEL BEARINGS

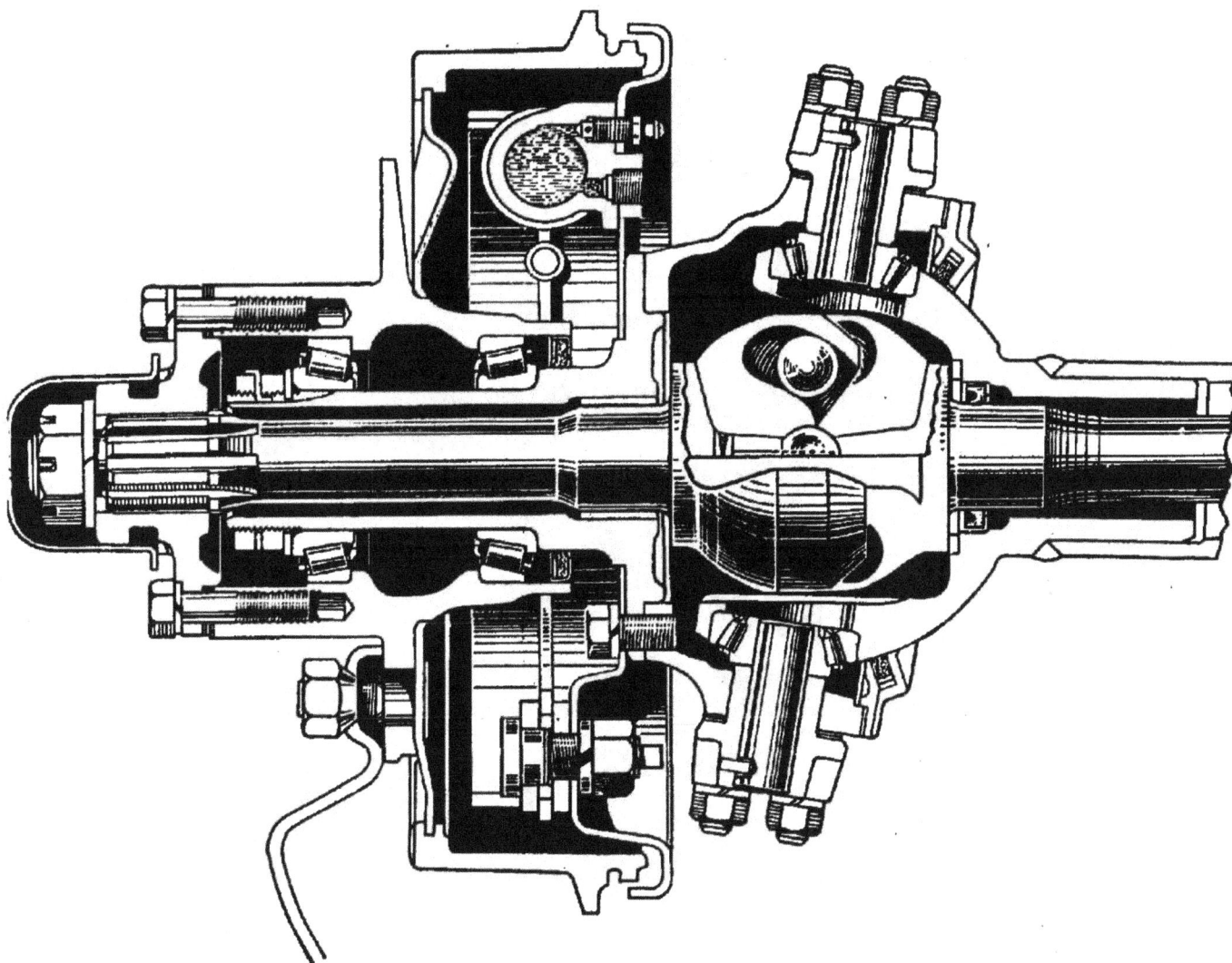

FIG. 1—FRONT WHEEL

The front and rear wheels are carried on two opposed tapered roller bearings. Bearings are adjustable for wear and their satisfactory operation and long life depends upon periodic attention and correct lubrication.

Wheel bearings cannot be checked for adjustment properly unless brakes are free from dragging on brake drums and are in fully released position.

Front Wheel Bearings

1. Raise front end of vehicle with jack so that tires clear the floor.
2. With hands test sidewise shake of the wheel. If bearings are correctly adjusted, shake of wheel will be just perceptible and wheel will turn freely with no drag. If bearing adjustment is too tight, the rollers may break or become overheated. Loose bearings may cause pounding.

If this test indicates adjustment is necessary, proceed as follows:

Adjustment

1. With wheels still on jack remove hub cap, axle shaft nut, washer and driving flange. Wheel bearing adjustment will then be accessible.
2. Bend lip of nut lock so that adjustment locknut and lock can be removed.
3. Tighten adjusting nut until wheel binds, at the same time rotating wheel to make sure all surfaces are in proper contact.
4. Then back off nut about ⅛ turn or more if necessary making sure wheel rotates freely.
5. Replace lock and do not fail to bend over locknut.
6. Check adjustment and reassemble driving flange. When front hub is completely assembled, test wheel shake before removing jack.

Rear Wheel Bearings

Raise wheel on which adjustment is to be made by placing jack under axle housing. Test wheel for loose bearing. If adjustment is necessary proceed as follows:

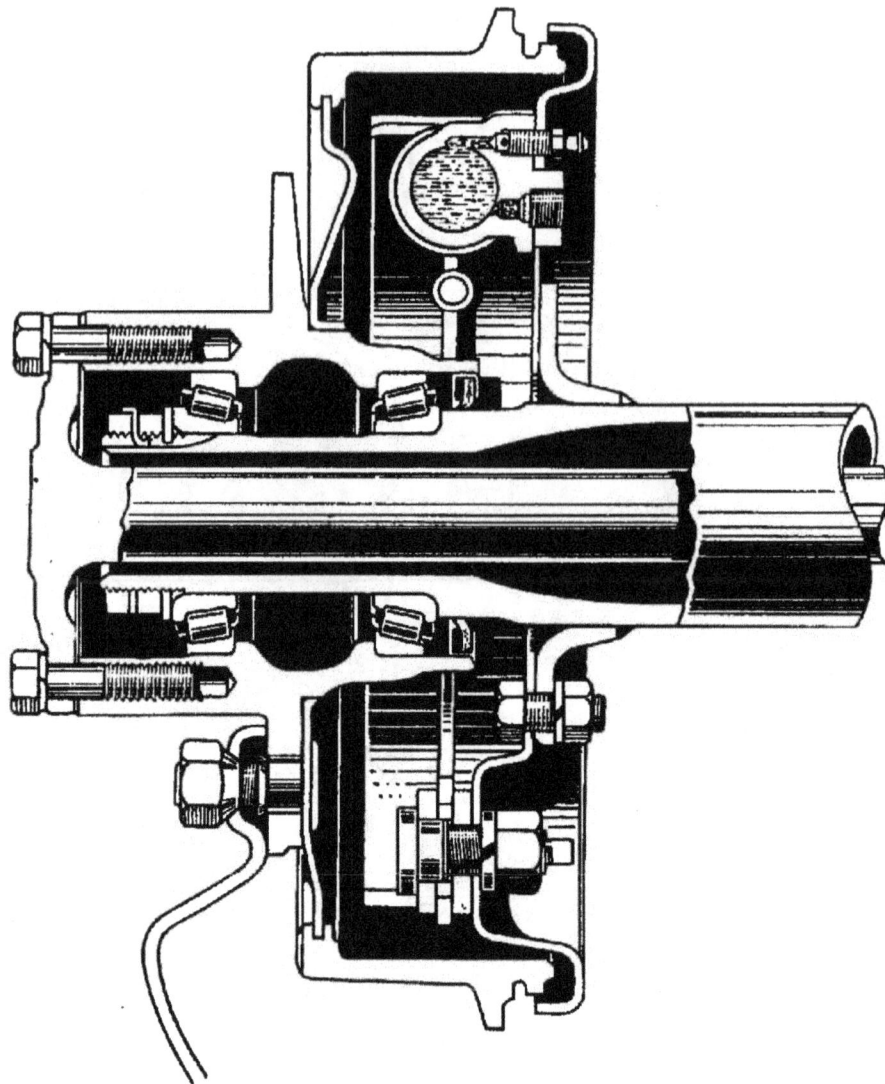

FIG. 1—REAR WHEEL

Adjustment

1. Remove axle shaft flange cap screws and axle shaft.

2. Bend lip of nut lock so that locknut can be removed.

3. Tighten inner adjusting nut until wheel binds, at the same time rotate wheel to make sure all surfaces are seating properly.

4. Back off nut ⅙ turn or more if necessary until wheel turns freely.

5. Replace nut lock and locknut and be sure to bend over lock.

6. Replace axle shaft with gasket and install cap screws.

Lubrication Wheel Hub Bearings

Under normal operating conditions the hub bearings require lubrication only every 6,000 miles when hubs and bearings should be removed and thoroughly washed in suitable cleaning fluid.

Bearings should be given more than a casual cleaning. Use a clean stiff brush and remove all particles of old lubricant from bearings and hubs.

After bearings are thoroughly cleaned inspect for pitted races and rollers, also check the hub oil seal.

Repack bearing cones and rollers with grease and reassemble hub in reverse order as that of dismantling, testing bearing adjustment as covered under "Adjustment"

When reinstalling hubs and drums the hubs with the right hand threaded studs are placed on the right hand side of vehicle. The left hand threaded studs are on the left hand side, viewing vehicle from the rear.

Brake Drum

The brake drums are attached to the wheel hubs by five serrated bolts. These bolts are also used for mounting the wheels to the hubs.

To remove a brake drum, drive out the serrated bolts and remove the drum from hub.

When placing drum on hub, make sure that the contacting surfaces are clean and flat. Line up holes in drum with those in hub and force drum over shoulder on hub. Insert five new serrated bolts through drum and hub and drive the bolts into place solidly. Place a round piece of stock in vise approximately the diameter of the head of the bolt and place hub and drum assembly over it so that it rests against head of the bolt then swedge bolt into countersunk section of hub with punch.

The runout of the face of the drum should be within .003". If runout is found to be greater than .003" it will be necessary to reset bolts to correct the condition.

Left hand hub bolts are identified with an "L" stamped on threaded end of bolt.

The left hand threaded nuts can be identified by a groove around the hexagon faces.

Hubs containing the left hand threaded hub bolts are installed on the left hand side of vehicle.

Tires

The most important factor of safe vehicle operation is systematic and correct tire maintenance. Tires must sustain the weight of a loaded vehicle, withstand more than ordinary rough service, provide maximum safety over all types of territory, and furnish the medium on which the vehicle can be maneuvered with ease.

Although there are other elements of tire service, inflation maintenance is the most important and in many instances the most neglected. Tire pressures should be consistently maintained for safe operation.

An under inflated tire is dangerous and too much flexing causes breakage of the fabric resulting in a failure. Over-inflation in time may cause a blow-out.

To remove the tire from a drop center rim, first deflate completely and then force the tire away from the rim throughout its entire circumference until the bead falls into the center of the wheel rim, then with a heavy screw driver or tire removing tool, placed across the wheel from the valve, remove one side of the tire at a time and remove inner tube. (See "Combat Wheels").

Installation of tire is made in the same manner by first dropping one side of the tire into the center of the rim and with tire tool spring bead over the wheel rim using care not to damage the inner tube.

When tightening the wheel stud nuts, alternately tighten opposite nuts to prevent wheel runout. After nuts have been tightened with the wheel jacked up, lower jack so wheel rests on the floor and retighten the nuts.

Combat Wheels

Combat wheels are identified by eight bolts holding together the two halves of the tire rim. When removing a tire, first remove the wheel and be sure to deflate the tire before removing the rim nuts. After removing the rim nuts, remove the outer rim then remove the tire after which remove the bead locking ring and tube from the tire. Mounting the tire is the reverse procedure. Do not put too much air in the tube when mounting. Combat wheel rim bolt and hub bolt torque reading 60-70 ft. lbs.

WHEEL SPECIFICATIONS

Wheels:

Make . . . . Kelsey-Hayes
Rim 16x4.00 Drop Center-16x4.50 Combat Wheels
Tires . . . . .16 x 6.00
Type . . . . . Mud and Snow non-directional Tread
Tire Pressure . . . . .30 lbs.

Bearings—F and R	Inner	Outer
Make . . . . .Timken		Timken
Cone and roller 18590		18590
Cup 18520		18520

STEERING

The stability and proper functioning of the steering system depends in a large measure upon correct alignment and a definite procedure for inspection of the steering system is recommended. In so doing, nothing is overlooked and the trouble is ascertained in the shortest possible time. It is suggested that the following sequence be used:

1. Equalize tire pressures and level car.
2. Inspect king pin and wheel bearing looseness.
3. Check wheel runout or wobble.
4. Test wheel balance.
5. Check for spring sag.
6. Inspect brakes and shock absorbers.
7. Check steering assembly and connecting rod.
8. Check caster.
9. Check toe-in.
10. Check toe-out on turns.
11. Check camber.
12. Check king pin inclination.
13. Check tracking of front and rear axle.
14. Check frame alignment.

The steering gear Fig. 1, is the cam and twin lever variable ratio type. The steering gear cam lever shaft is serrated for attachment to the steering pitman arm. The gear case is attached to the inside of the left frame side member by three bolts.

The cam thrust is taken at top and bottom by ball bearings which are adjustable through shims No. 13 at the upper housing cover No. 12.

When making adjustments free the steering gear of all load by disconnecting the steering connecting rod from the steering arm, loosen instrument panel bracket and steering gear frame bolts to allow the steering post to align itself.

Do not tighten the steering gear to dampen out steering troubles, adjust the steering only to remove play within the steering gear.

Adjustment of Ball Thrust Bearings on Cam

Adjust to a barely perceptible drag but allow the steering wheel to turn freely, with thumb and forefinger lightly gripping the rim.

Before making this adjustment, loosen the housing side cover adjusting screw No. 19 to free the pins in the cam groove.

To adjust, remove cap screws and move up the housing cover No. 12 to permit the removal of shims No. 13. Shims are of .002", .003", and .010" thickness.

Clip and remove a thin shim or more if required. Install cap screws and tighten. Test adjustment and if necessary remove or replace shims until adjustment is correct.

Adjustment of Tapered Pins in Cam Groove

Adjust so that a very slight drag is felt through the mid position when turning the steering wheel slowly from one extreme position to the other.

Backlash of the pins in the groove shows up as end play of lever shaft, also as backlash of steering at ball on steering arm.

Note that the groove is purposely cut shallow in the straight ahead driving position for each pin. Fig. 2. This feature permits a close adjustment for normal straight ahead driving thereby avoiding swaying in the road and also permits take-up of back-lash at this point after wear occurs without causing a bind elsewhere.

Adjust within the high range through the mid position of pin travel. Do not adjust the positions off "straight ahead". Backlash in turn positions is not objectionable.

Removal of Steering Gear from Chassis

To remove steering gear assembly from chassis, the following procedure should be followed:

1. Remove left front fender.
2. Remove horn button and steering wheel.
3. Remove steering post bracket at instrument board.
4. Remove steering post cover plates on toe board.
5. Remove horn wire contact brush, No. 37, Fig. 1.
6. Remove connecting rod at Pitman arm ball.
7. Remove three bolts holding steering gear housing to frame side rail.
8. Remove steering post by bringing it down through floor boards and over outside of the frame.

The installation of the steering gear assembly would be the reverse of the above operations. Frame bolts, torque wrench reading, 36-40 ft. lbs.

Disassembly of Steering Gear

First remove pitman arm No. 21, Fig. 1 with puller. Loosen lock nut No. 20 and unscrew adjusting screw No. 19 a few turns. Remove side cover screws and washers and remove side cover No. 18 with gasket. This will permit removal of lever shaft No. 16.

Remove upper cover plate screws and remove from housing the cam, wheel tube and bearing assembly.

When upper cup or upper cover plate requires replacement, the contact ring on wheel tube must be removed. To do this unsolder horn cable from ring and pull cable from tube, mark on wheel tube the location of ring and then press ring off of tube.

Inspect cam threads for wear, chipping and scoring, also the ball races on the cam ends and the separate ball cups. Existence of any of these conditions indicate necessity for replacement.

Inspect taper studs of lever shaft for flat spots and chipping. In the case of either, replacement is usually advisable. Inspect lever shaft for wear and test fit of shaft in bushing. Inspect condition of oil seal at outer end of lever shaft and the bearing in top end of jacket tube.

Assembling Steering Gear

Reassemble all parts to wheel tube in reverse order of disassembly and flatten cable and solder to ring. Assemble cam, wheel tube and bearing assembly in housing, seating well the lower bearing ball cup in the housing.

With adjusting shims in place, assemble upper cover plate with pin on top side of housing and adjust cam bearings.

Assemble lever shaft in housing and with gasket in place assemble side cover and make adjustment for a minimum backlash of studs in cam groove.

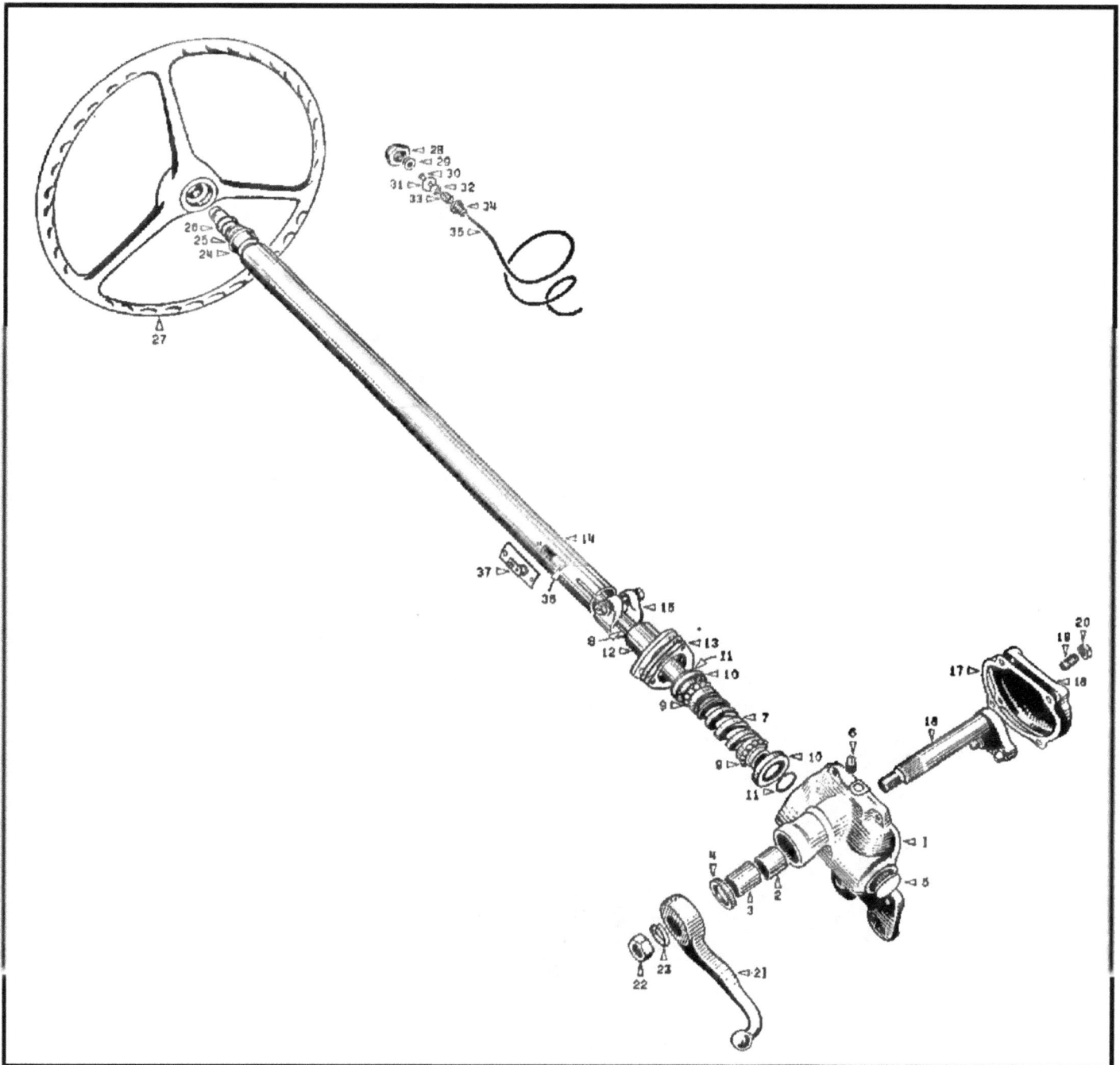

FIG. 1—STEERING GEAR

No.	Willys Part No.	Ford Part No.	Name	No.	Willys Part No.	Ford Part No.	Name
1	A-740	GPW-3548	Housing Assembly	20	52925	33927-S	Side Adjusting Screw Lock Nut
2	639090	GPW-3587	Housing Bushing—Inner	21	A-1116	GPW-3590	Steering Gear Arm
3	639091	GPW-3576	Housing Bushing—Outer	22	639115	356077-S8	Steering Gear Arm Nut
4	639095	GPW-3591	Housing Oil Seal	23	5038		Steering Gear Arm Nut Lockwasher
5	51091	74121-S	Housing Lower End Plug	24	639190	GPW-3517	Steering Column Bearing Assembly
6	5085	358064-S	Housing Oil Filler Plug	25	639192	GPW-3518	Steering Column Bearing Spring Seat
7-8	A-742	GPW-3524	Cam and Wheel Tube Assembly	26	639191	GPW-3520	Steering Column Bearing Spring
9	639104	GPW-3671	Cam Bearing Balls	27	A-635	GPW-3600	Steering Wheel
10	639102	GPW-3552	Ball Cup—Upper and Lower	28	A-633	GPW-3655	Steering Wheel and Horn Button Nut
11	639103	GPW-3589	Ball Cup Retaining Ring—Upper and	29	A-634	GPW-3627	Horn Button
			Lower	30	638886	GPW-3630	Horn Cable Upper Terminal
12	A-1760	GPW-3568	Upper Housing Cover	31	A-750	GPW-3631	Contact Washer
13	639108	GPW-3593	Adjusting Shims	32	A-751	GPW-3635	Insulating Ferrule
14	A-1199	GPW-3509	Steering Column and Bearing Assembly	33	638884	GPW-3626	Horn Button Spring
15	A-635	GPW-3506	Steering Column Clamp Assembly	34	638885	GPW-3646	Horn Button Spring Cup
16	A-745		Lever Shaft Assembly	35	A-752	GPW-14171	Horn Cable Assembly
17	639119	GPW-3581	Side Cover Gasket	36	A-747	GPW-3652	Horn Wire Contact Ring Assembly
18	639117	GPW-3580	Side Cover	37	A-302	GPW-13836	Horn Wire Contact Brush Assembly
19	639118	GPW-3577	Side Adjusting Screw				

When assembling upper bearing spring No. 26, and spring seat No. 25 in jacket tube make sure that spring seat is positioned correctly. It must be placed with the lengthwise flange down against bearing and not up inside of spring coil.

Install pitman arm No. 21 to lever shaft No. 16 so that the line across the face of arm and end of shaft correspond, with the ball end down. Install lockwasher No. 23 and nut No. 22.

Install steering gear assembly in chassis in the reverse order in which it was removed.

When installing the steering wheel the steering gear should be in its mid position when the front wheels are in the straight ahead position. To check, turn the steering wheel as far to the right as possible then rotate the wheel in the opposite direction as far as possible and note the total number of turns. Turn the wheel back just one half of this total movement thus placing the gear in mid position at which point the front wheels should be straight ahead. The steering wheel spoke with moulded trade mark on underside will point down toward drivers seat and in line with the steering post. If not it will be necessary to remove the steering wheel and shift it on the serrations of the shaft.

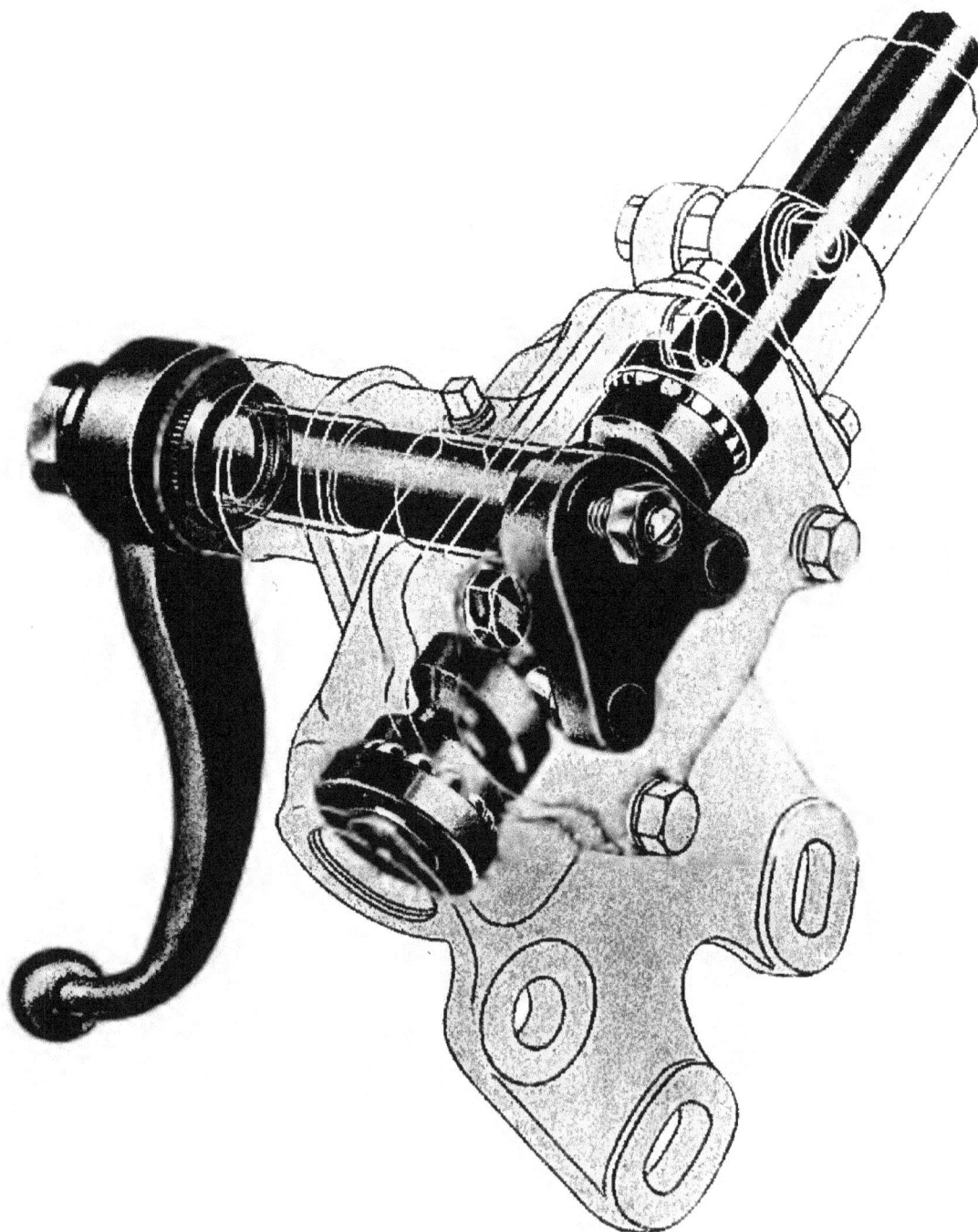

FIG. 2—SECTIONAL VIEW OF STEERING

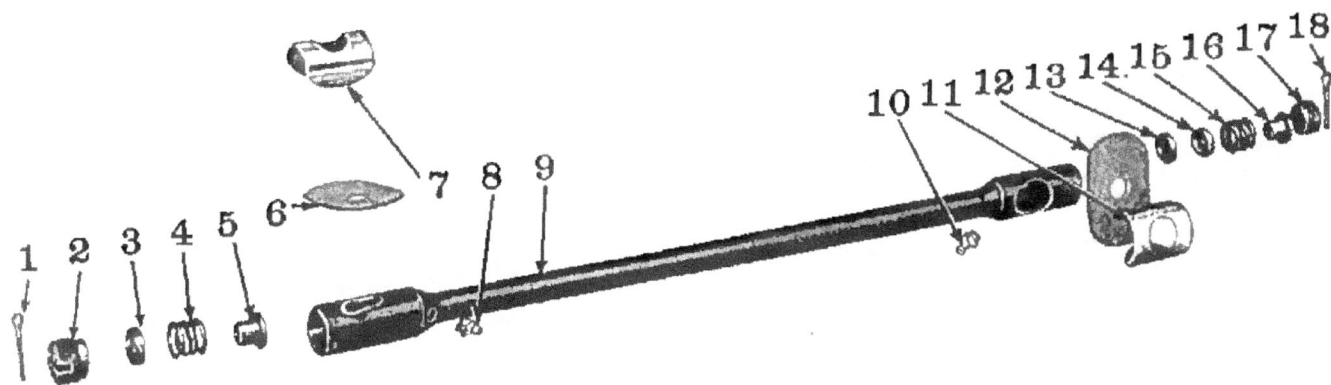

FIG. 3—STEERING CONNECTING ROD

No.	Willys Part No.	Ford Part No.	Name	No.	Willys Part No.	Ford Part No.	Name
1	5134	72089-S	Steering Connecting Rod Cotter Pin—Front	10	392909	353047-S7-8	Hydraulic Straight Grease Fitting
2	630756	GPW-3323	Steering Connecting Rod Adjusting Plug—Large	11	A-623	GPW-3336	Steering Connecting Rod Dust Cover Shield
3	630755	GPW-3320	Steering Connecting Rod Ball Seat	12	A-622	GPW-3332	Steering Connecting Rod Dust Cover
4	630754	GPW-3327	Steering Connecting Rod Spring	13	630755	GPW-3320	Steering Connecting Rod Ball Seat
5	630753	GPW-3326	Steering Connecting Rod Safety Plug	14	630755	GPW-3320	Steering Connecting Rod Ball Seat
6	A-622	GPW-3332-A2	Steering Connecting Rod Dust Cover	15	630754	GPW-3327	Steering Connecting Rod Spring
7	A-623	GPW-3336	Steering Connecting Rod Dust Cover Shield	16	630753	GPW-3326	Steering Connecting Rod Safety Plug
8	392909	353027-S7-8	Hydraulic Straight Grease Fitting	17	630757	GPW-3328	Steering Connecting Rod Adjusting Plug—Small
9	A-619	GPW-3304	Steering Connecting Rod Assembly	18	5134	72089-S	Steering Connecting Rod Cotter Pin—Rear

Steering Connecting Rod

The steering connecting rod is the ball and socket type. At front or axle end, the spring and spacer are assembled between rod (bottom of socket) and ball seat while at the steering gear end, spring and spacer are between ball seat and end plug. See Fig. 3.

When removing springs and seats for any reason make sure they are reassembled as above because this method of assembly relieves road shock from the steering gear in both directions. To adjust ball joint at axle, screw in plug firmly against the ball, then back off one half turn and lock with new cotter pin inserted through hole in tube and slot in adjusting plug.

To adjust ball joint at steering Pitman Arm, screw in end plug firmly against the ball, then back off one full turn and lock with new cotter pin inserted through hole in tube and slot in adjusting plug.

This will give the proper spring tension and avoid any tightness when swinging the wheels from maximum left to right turn.

Ball joints must be tight enough to prevent end play and yet loose enough to allow free movement.

Tie Rod

The tie rods, No. 11 and 14, Fig. 4 are of three piece construction consisting of rod and two ball and socket end assemblies. Ball and socket end assemblies are threaded into rod and locked with clamps around each end of tie rod. Right and left hand threads on tie rod end assemblies provide for toe-in adjustment without removing the tie rod ends from steering arms.

The length of the left hand tie rod No. 14 center to center of ball joint is 17¹¹⁄₃₂", the right hand tie rod No. 11 is 24¼" center to center.

When wear takes place on tie rod end ball and socket, it will be necessary to replace the ball and socket assembly and rubber seal.

Front Wheel Alignment

Proper alignment of front wheels must be maintained in order to insure ease of steering and satisfactory tire life. Most important factors of front wheel alignment are wheel camber, axle caster and wheel toe-in.

These points should be checked at regular intervals particularly where the front axle has been subjected to heavy impact. When checking wheel alignment, it is important that wheel bearings and knuckle bearings be in proper adjustment. Loose bearings will affect reading of instruments when checking camber, knuckle pin inclination and toe-in.

Wheel toe-in is the distance the wheels are closer together at the front than at the rear.

Wheel camber is the amount wheels incline outward at the top from a vertical position.

Front axle caster is the amount in degrees that the steering knuckle pins are tilted toward the front or rear of the vehicle. Positive caster is inclination of top of knuckle pin toward rear of vehicle. Zero caster is vertical position of pin. Negative or reverse caster is the inclination of top of pin towards the front of the vehicle.

Front Wheel Toe-in

Toe-in Fig. 5, is the amount which wheels point inward at front and is necessary to offset the effect of camber.

Toe-in is usually measured at edge of rim, flange or at tire centers with wheels in straight ahead position, however in view of the tread being the same, front and rear, a straight edge or rope can be used.

It is highly important that the toe-in be checked regularly and if found to be excessively out of adjustment, correction should be made immediately.

To Adjust Toe-in

1. Set tie rod end of steering bell crank at right angles with front axle.

2. Place a straight edge or rope against the left rear wheel and left front wheel to determine if wheel is in straight ahead position. If the rear of tire on front wheel does not touch straight edge, it will be necessary to adjust the tie rod by loosening clamps on each end and turning the rod in a clockwise direction until the tire touches the straight edge both front and rear. If the front of the tire does not strike the straight edge, it will be necessary to lengthen the tie rod by turning the rod in a counter-clockwise direction.

3. Check the right hand side in the same manner adjusting the tie rod if necessary, making sure that the bell crank remains at right angles to the axle.

4. Set the toe-in to $\frac{3}{64}$"-$\frac{3}{32}$" by shortening each tie rod approximately one half turn.

Front Wheel Camber

The purpose of camber Fig. 6 is to more nearly place the weight of the vehicle over the center of the steering knuckle pins and facilitate easy steering.

The result of excessive camber is irregular wear of tires on outside shoulders and is usually caused by bent axle parts.

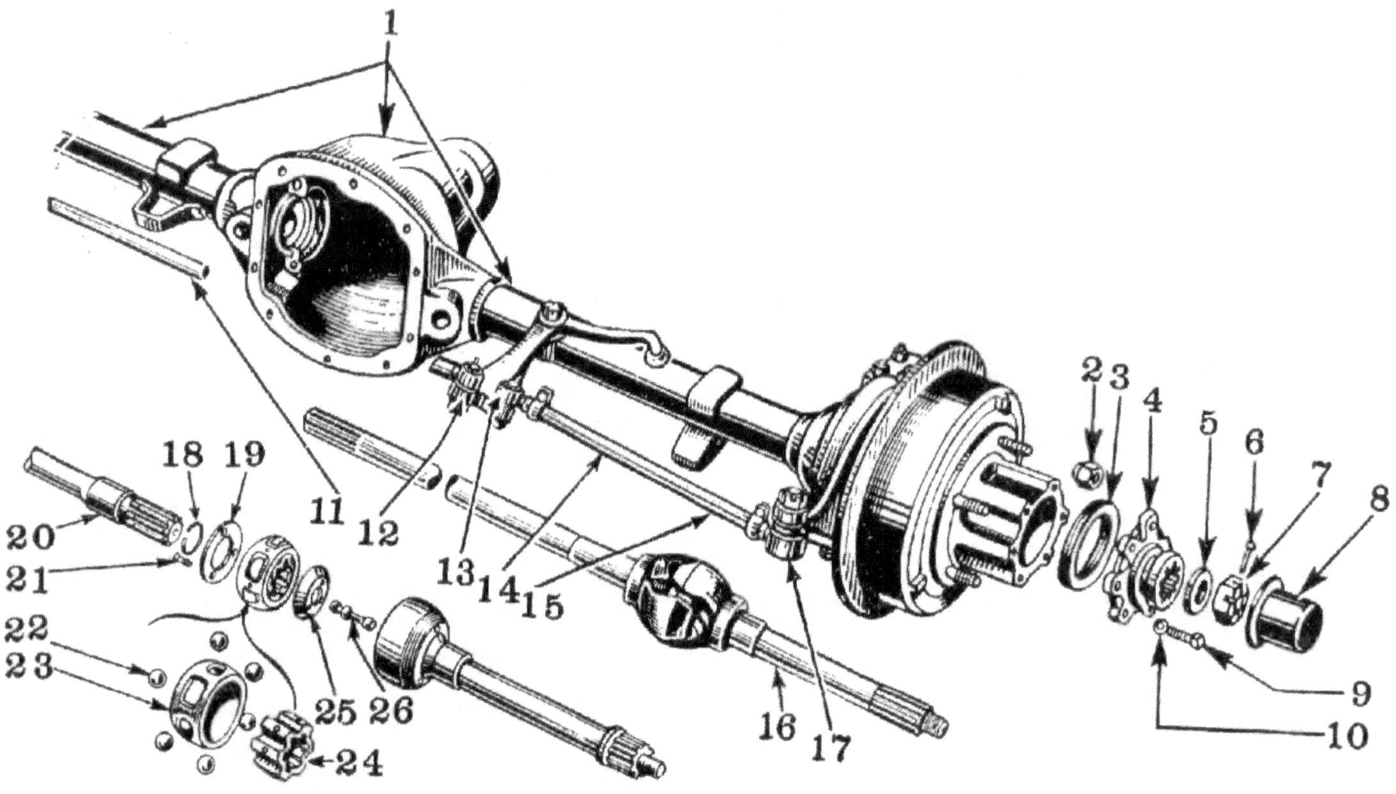

FIG. 4—FRONT AXLE ASSEMBLY

No.	Willys Part No.	Ford Part No.	Name		No.	Willys Part No.	Ford Part No.	Name	
1	A-1703	GPW-3074	Gear Carrier Housing and Tube Assembly		17	A-847	GP-3290	Tie Rod Socket Assembly—Left	
2	A-476	GP-1012	Wheel Hub Bolt Nut R.H. Thread (Ford GP-1013; Willys A-475 L.H. Thd.)		18	A-1726	GP-3216	Axle Shaft Retainer Snap Ring	
3	A-862	GP-3208-A	Universal Joint Adjusting Shims		19	A-1724	GP-3217	Axle Shaft Retainer	
4	A-868	GP-3204	Axle Shaft Drive Flange		20	A-1729	GPW-3017-A	Axle Inner Shaft—Left (Ford GPW-3016-A; Willys A-1727 Right Inner)	
5	636570	356504-S	Axle Shaft Nut Washer		21	A-1725	24622-S	Axle Shaft Retainer Screw	
6	5397	72071-S	Axle Shaft Nut Cotter Pin		22	A-1721	358074-S	Axle Shaft Universal Joint Ball	Rzeppa Universal Joint
7	636569	356126-S	Axle Shaft Nut		23	A-1719	GP-3215	Axle Shaft Universal Joint Cage	
8	A-869	GP-1139	Hub Cap		24	A-1720	GP-3221-A	Axle Shaft Universal Joint Inner Race	
9	A-760	GP-1110	Axle Shaft Drive Flange Cap Screw		25	A-1722	GP-3219	Axle Shaft Universal Joint Pilot	
10	5010	34807-S	Axle Shaft Drive Flange Screw Lockwasher		26	A-1723	GP-3218	Axle Shaft Universal Joint Pilot Pin	
11	A-1705	GPW-3281	Tie Rod Tube—Right						
12	A-1211	GPW-3131	Drag Link Bell Crank						
13	A-838	GP-3289	Tie Rod Socket Assembly—Right						
14	A-1709	GPW-3282	Tie Rod Tube—Left						
15	A-1708	GPW-3279	Tie Rod Assembly—Left						
16	A-809	GPW-3206-A2	Axle Shaft and Universal Joint Assembly R.H. (Bendix) (Ford GPW-3207-A; Willys A-810 Left Hand)						

The result of negative or reverse camber, if excessive, will be hard steering and possibly a wandering condition. Tires will also wear on inside shoulders. Negative camber is usually caused by excessive wear or looseness of front wheel bearings, axle parts or the result of a sagging axle.

Result of unequal camber may be any or a combination of the following conditions—unstable steering, wandering, kick-back or road shock, shimmy or excessive tire wear. The cause of unequal camber is usually a bent steering knuckle or axle center.

Correct wheel camber is 1½° and is set in the axle at the time of manufacture and cannot be altered by any adjustment. It is important that the camber be the same in both front wheels. Excessive heating of these parts to facilitate straightening destroys the heat treatment given them at the factory. Cold straightening of bad bends may cause a fracture of the steel and is unsafe. Replacement with new parts is recommended rather than any straightening of damaged parts.

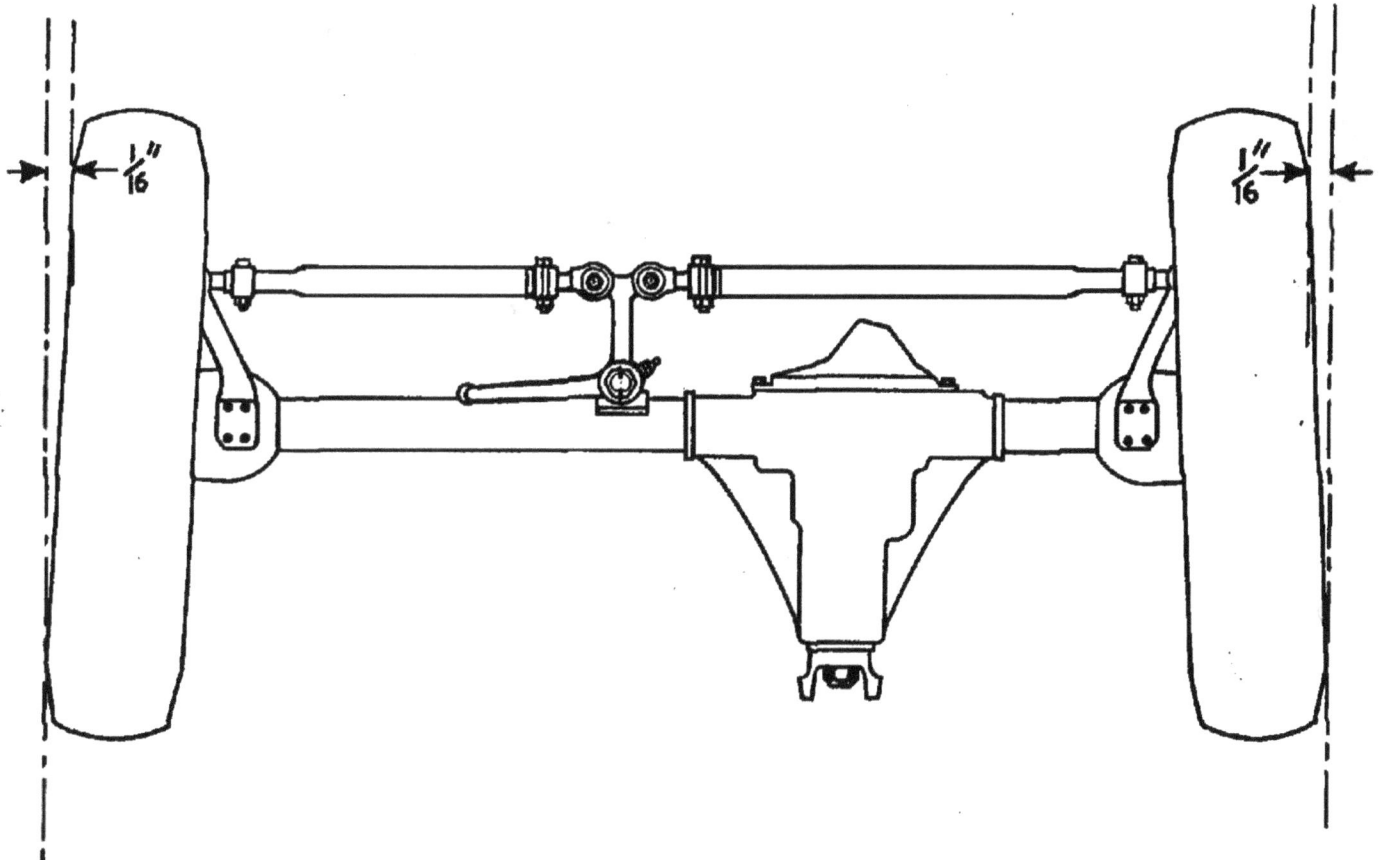

FIG. 5—FRONT WHEEL TOE-IN

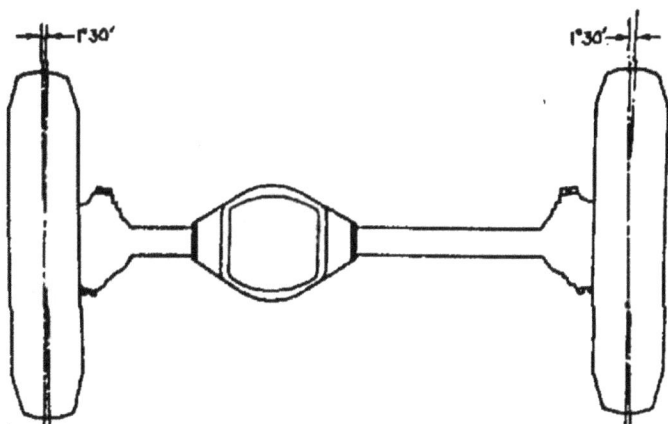

FIG. 6—WHEEL CAMBER

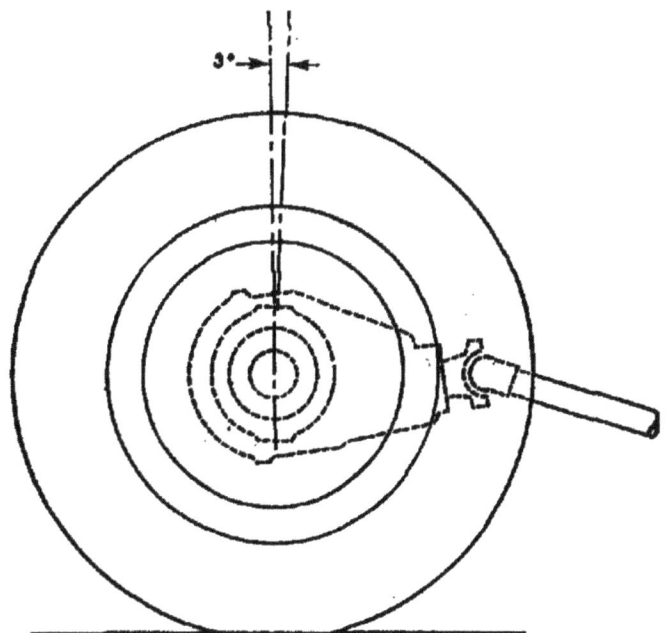

FIG. 7—AXLE CASTER

Axle Caster

The purpose of caster Fig. 7 is to provide steering stability which will keep front wheels in a straight ahead position and to assist in bringing wheels out of a turn on a curve.

The result of no caster is wandering or the vehicle will not come out of a turn normally.

No adjustment is provided for this angle but if checked with a suitable gauge and found to be incorrect, it should be investigated and if not excessive, correction made by the use of axle wedges or bending the axle cold.

If the camber and toe-in are correct and it is known that the axle is not twisted, a satisfactory check can be made by testing vehicle on the road. Before road testing make sure all tires are properly inflated, being particularly careful that both front tires are inflated to exactly the same pressure.

If vehicle turns easily to both sides but is hard to straighten out, this indicates insufficient caster for easy handling of vehicle.

Front Wheel Turning Angle

When the front wheels are turned the inside wheel on the turn travels in a smaller arc than the outside wheel; therefore it is necessary for the wheels to toe out. This change in wheel alignment is obtained through the length and angularity of the steering knuckle arms in relation to the front axle. When the wheels are turned so the inside wheel is on an angle of 20° as shown by "I" in Fig. 8, the outer wheel angle "O" should be 19°45′. The left steering knuckle arm controls the relationship of the front wheels on a left turn and the right arm controls the relation on a right turn. If a steering arm should be accidently bent it can be straightened cold if the bend is not excessive, otherwise the arm should be replaced.

Steering Bell Crank

The bell crank is a drop forging, heat treated for strength with removable ball ends, steering connecting rod and tie rods.

The bell crank is mounted on the front axle and swivels on two needle bearings.

If bell crank becomes damaged or bent do not attempt to heat and straighten, straighten it cold or install new parts.

The bell crank shaft is removable from axle by driving out tapered lock pin, driving pin out toward left front wheel.

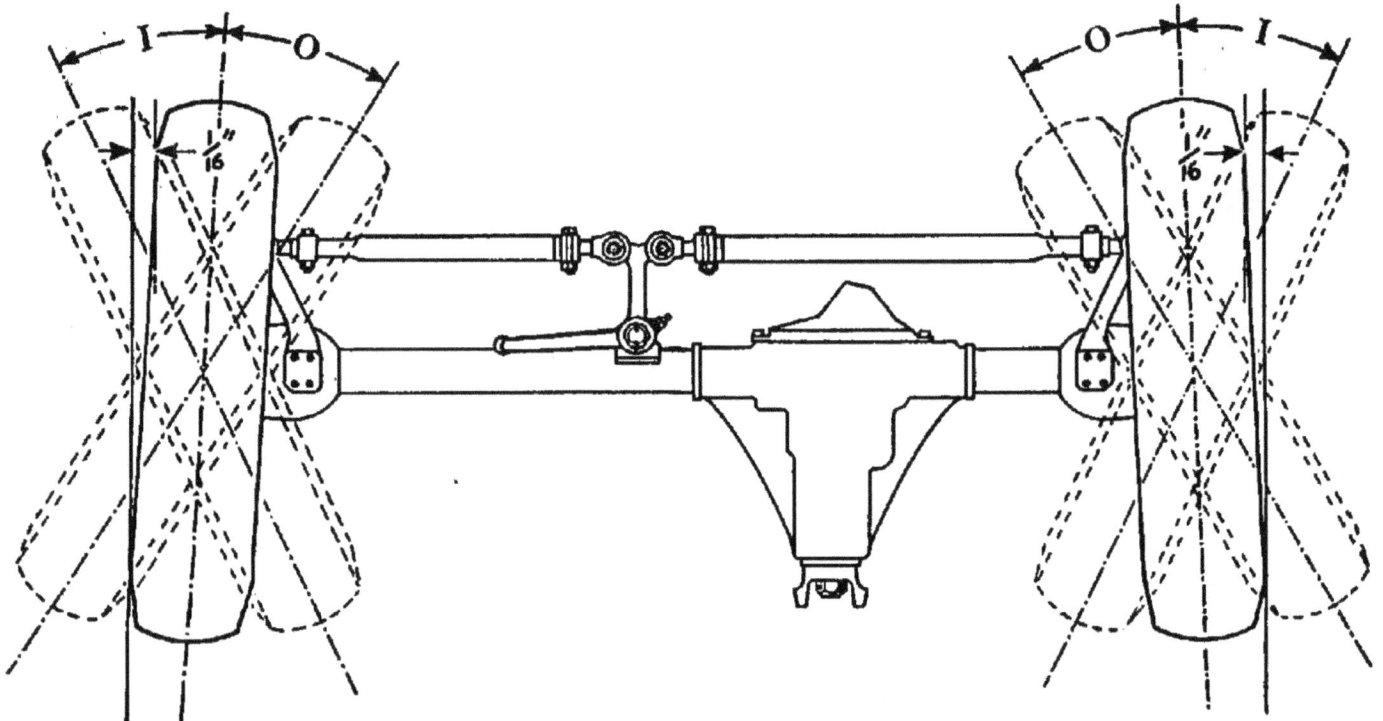

FIG. 8—FRONT WHEEL TURNING ANGLE

STEERING TROUBLES AND REMEDIES

SYMPTOMS	PROBABLE REMEDY

Hard Steering

Lack of Lubrication.........................Lubricate All Connections
Tie Rod Ends Worn..........................Replace
Connecting Rod Ball Joints Tight.............Adjust
Cross Shaft Improperly Adjusted............Adjust
Steering Gear Parts Worn...................Replace
Front Axle Trouble.........................See "Front Axle" Section

Steering Loose

Tie Rod Ends Worn.........................Replace
Connecting Rod Ball Sockets Worn...........Replace
Steering Gear Parts Worn...................Replace
Steering Gear Improperly Adjusted...........Adjust

Road Shock.........................Steering Connecting Rod too tight; Axle Spring Clips loose; Wheel Bearings loose; poor Shock Absorber control.

Turning Radius

Short one side.........................Center Bolt in spring sheered off, Axle shifted, Steering Arm bent, Steering Arm not properly located on Steering Gear.

STEERING SPECIFICATIONS

Steering Gear:

Make.........................Ross
Type..............Cam and Twin Pin Lever
Model.........................T-12
Ratio..........Variable Ratio, 14-12-14 to 1
Wheel..........3 spoke—17¼", Safety type

Bearings:

Cam Upper.........................Ball
Cam Lower.........................Ball
Levershaft.........................Bushing
Steering Column Upper...............Ball

Lever Shaft:

Clearance to Bushing...........0005"-.0025"
End Play.........................000"
Lash at Cam (Straight ahead)...............
.................Slight drag over high point

Steering Connecting Rod:

Make.................Columbus Auto Parts
Type.........................Spring loaded
Adjustment...................Threaded Plug

Steering Geometry:

Toe-in.........................$\frac{3}{64}$"-$\frac{3}{32}$"
Camber.........................1½°
Caster.........................3°
Toe out
  Inside wheel.........................20°
  Outside wheel...................19°45'

FRAME

The frame is the structural center of any vehicle, for in addition to carrying the load, it provides and maintains correct relationship between other units to assure their normal functioning.

The frame is of rugged design and constructed of heavy channel steel side rails and cross members. Braces and brackets are used to maintain the proper longitudinal position of the side rails relative to each other, and at the same time provide additional resistance to torsional strains. Due to this rugged design the frame requires very little attention to maintain its dependability.

Vehicles which have been in an accident of any nature which may result in a swayed or sprung frame, should always be carefully checked for proper frame alignment in addition to steering geometry and axle alignment.

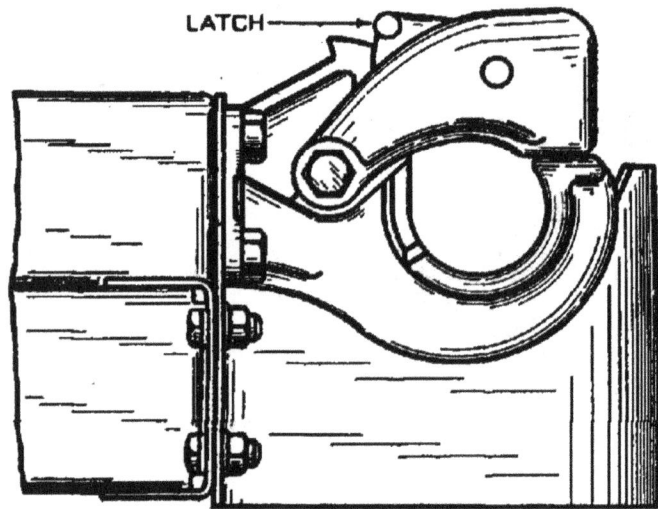

FIG. 1—PINTLE HOOK

A pintle hook for towing is provided on the rear frame crossmember. To open the hook lift up on the safety latch.

Checking Frame Alignment

When checking a frame for alignment Fig. 2, the most efficient method is "X" checking from given points on each side rail.

The most convenient way to check frame alignment, particularly when body is on chassis is by marking on floor all points from which measurements should be taken.

Select a space on the floor which is comparatively level. If cement floor, clean so that chalk marks will appear underneath the points of frame to be checked. If a wooden floor, it is advisable to lay a sheet of paper underneath the vehicle and tack in place, dropping a plumb-bob from each point indicated in Fig. 2, mark flooring directly underneath plumb-bob. Satisfactory checking depends upon the accuracy of marks with relation to the frame.

To reach points shown that have been marked, have vehicle carefully moved away from layout on floor, and proceed as directed in the following paragraphs:

1. Check frame width at front and rear ends using corresponding marks on floor. If widths correspond to specifications given draw center line full length of vehicle half way between marks indicating front and rear widths. If frame width is not correct lay out center line as follows:

   If center line cannot be laid out from checking points of ends of frame, it can be drawn through intersection of any two pairs of equal diagonals. If the extreme front end of frame is damaged, center of front end of frame can be located from point exactly midway between radiator support bolts.

2. With the center line properly laid out, measure distance from it to opposite point marked over entire length of chassis. If frame is in proper alignment, measurement should not vary more then ⅛" at any station.

3. To locate point at which frame is sprung measure diagonals marked "AB" "BC" "CD". If the diagonals in each pair are within ⅛", that part of the frame included between points of measurements may be considered in satisfactory alignment. These diagonals should intersect within ⅛" of center line. Any variations of more than ⅛" indicates misalignment. If the measurements do not agree within the above limits, it means that correction will have to be made between those measured points that are not equal.

Straightening Frame

In the case where the bending or twisting of the frame is not excessive, the frame may straighten. This should be done cold, excessive heat applied to the frame might change the structure of the metal and weaken the frame. For this reason it is recommended that badly damaged frame parts be replaced.

Front Axle Alignment

After it has been determined that the frame is properly aligned, the front axle alignment with frame can be checked as directed below.

The front axle is square with the frame if the distance between the front and rear axle is the same on both sides, and the distance from center of the upper spring bushing to front axle on both sides are equal.

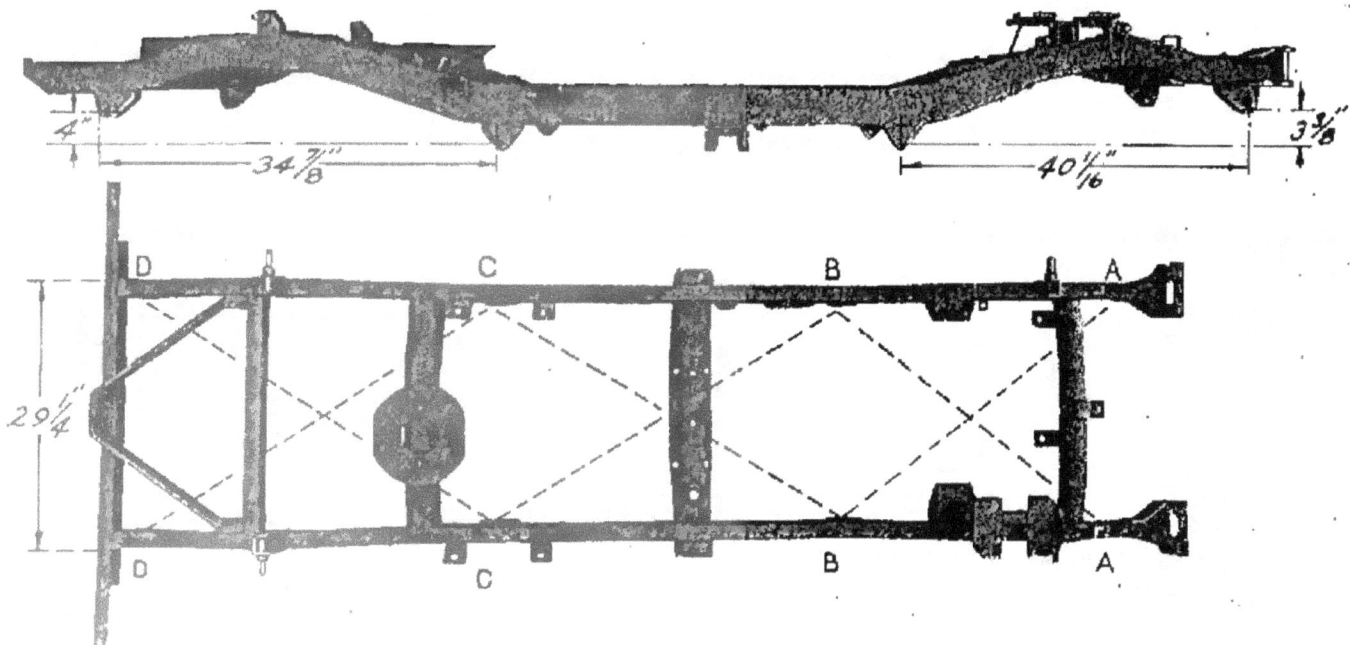

FIG. 2—FRAME ASSEMBLY

FRAME SPECIFICATIONS

Frame . . . . SAE 1025
    Depth Maximum . . . . 4.186".
    Thickness Maximum . . . . .083"-.093"
    Flange Width . . . . 1¾"
    Length . . . . 122¾"

Width
    Front . . . . 29¼"
    Rear . . . . 29¼"

Number Cross Members . . . . 5-"K" member at Rear
Weight . . . . 140 lbs.

Wheel Base . . . . 80"

Tread
    Front . . . . 48¼"-with combat wheels 49"
    Rear . . . . 48¼"-with combat wheels 49"

SPRINGS

The specially designed springs used on this vehicle are constructed of chromium alloy to stand the severe service to which they may be subjected.

Front Springs

The front springs Fig. 1 are the semi-elliptic type, 36¼" long and 1¾" wide. There are eight leaves in each spring, seven leaves being of the parabolic shape with No. 2 leaf military wrapped over the eye of No. 1 leaf. The ends of the leaves are turned down to eliminate squeaks. Each spring is equipped with four rebound clips 1¼" wide.

The front springs appear to be identical in construction, nevertheless, they are different in load carrying ability.

The left spring requires a load of 525 lbs. for a 5/16" camber. The right front spring requires a load of 390 lbs. for a 5/16" camber. This difference is required due to the extra weight on left side of vehicle. The left spring can be identified by letter "L" painted on lower side at front on second leaf.

The front end of the front springs are shackled, using the "U" type shackle with threaded core bushing. The rear end of the spring is bronze bushed and is pivoted by a pivot bolt in the bracket on the frame. A torque reaction spring stabilizes the torque of the front axle.

The spring saddles on axle are welded in place to the underside of axle housing and springs are held in that position through "U" bolts, using the center spring bolt inserted in spring saddle to prevent the shifting of the axle.

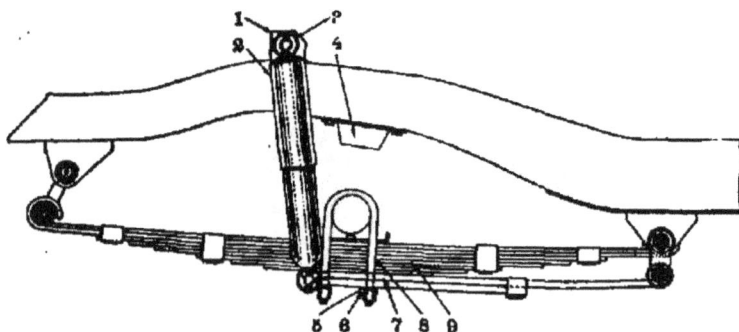

FIG. 1—LEFT FRONT SPRING AND
SHOCK ABSORBER

No.	Willys Part No.	Ford Part No.	Name
1	A-1204		Front Shock Absorber Bracket and Shaft Assembly Left (A-1205 Right)
2	A-169	GPW-18045	Shock Absorber Assembly
3	637936	GPW-18060	Shock Absorber Mounting Pin Bushing (Rubber)
4	A-481	GPW-5783	Axle Bumper
5	5938	34848-S	Spring Clip Nut Lockwasher
6	339372	GPW-5456	Front Spring Clip Nut
7	A-6066	GPW-5588	Torque Reaction Spring Assembly
8	A-575	GPW-5705	Front Axle to Spring Clip, Left
9	A-612	GPW-5311	Front Spring—Left (Ford GPW-5310; Willys A-613 Right)

Rear Springs

The rear springs Fig. 2 are semi-elliptic, 42" long, 1¾" wide, 9 leaves with 4 rebound clips 1¼" wide. The spring leaves are the parabolic type with No. 2 leaf military wrapped around eye ends of No. 1 leaf. The ends of each leaf being turned down to eliminate squeaking.

The front end of the rear spring is bronze bushed and is pivoted by a pivot bolt at frame bracket, flexible "U" shackles are used at the rear.

The spring saddles are welded to the underside of rear axle housing and the center spring bolt is used to prevent shifting of the axle. The spring is held in position by two "U" bolts over the axle.

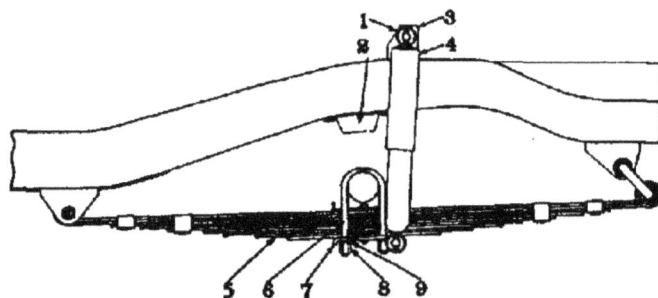

FIG. 2—REAR SPRING AND SHOCK ABSORBER

No.	Willys Part No.	Ford Part No.	Name
1	637936	GPW-18060	Shock Absorber Mounting Pin Bushing (Rubber)
2	A-481	GPW-5783	Axle Bumper
3	A-484		Rear Shock Absorber Bracket and Shaft Assembly (A-485—Right)
4	A-170	GPW-18080	Shock Absorber Assembly—Rear
5	A-614	GPW-5560	Rear Spring
6	A-575	GPW-5705	Rear Axle to Spring Clip
7	A-571	GPW-5460	Rear Spring Clip Plate and Shaft Assembly (Ford GPW-5459; Willys A-572 Right)
8	339372	GPW-5456	Rear Spring Clip Nut
9	5938	34848-S	Rear Spring Clip Nut Lockwasher

Spring Shackles and Pivot Bolts

The spring shackles are of the "U" type, Fig. 3 with threaded core bushings using right and left hand threads, depending at which position they are to be used in the chassis.

The bushings are anchored solidly in frame bracket and spring eyes and the oscillation taken between the threads of the "U" shackle and the inner threads of the bushing. The lubrication of the shackle bushings is very important, and should not be neglected, or excessive wear of the bushings and "U" shackles will occur.

There are six bushings used with right hand threads and two with left hand threads. The right hand threaded type bushings have plain hexagon head. The left hand threaded bushings have a groove around the head, Fig. 4.

The two left hand threaded "U" shackles can be identified by a forged boss on the lower shank of the shackle identifying the left hand thread Fig. 4. These two left hand threaded "U" shackles are used at the left front spring and the right rear spring, with the left hand threaded end at the spring eyes.

The "U" shackles are installed so that the bushing hexagon heads are to the outside of the frame. When making installation of a new "U" shackle or shackle bushing the following procedure should be followed.

Install shackle grease seal and retainer over threaded end of shackle up to the shoulder. Insert new shackle through frame bracket and eye of spring. Holding "U" shackle tightly against frame, start upper bushing on shackle, care being taken when it enters the thread in the frame that it is not cross threaded. Screw bushing on shackle about halfway, and then start lower bushing holding shackle tightly against spring eye and thread bushing in approximately halfway, then alternating from top bushing to lower bushing turn them in until the head of the bushing is snugly against the frame bracket, and the bushing in spring eye is $\frac{1}{32}''$ away from spring measured from inside of hexagon head to spring.

Lubricate the bushings with high pressure lubricant and then try the flex of the shackle, which should be free. If shackle is tight it will be detrimental to the bushings as well as to the spring and it will be necessary to rethread the bushings on shackle.

Remove and Replace Spring

To remove a spring raise the vehicle, then place a stand jack under frame side rail, adjusted to a distance so that the load is relieved on the spring and yet the wheels still rest on the floor, remove the four "U" bolt nuts and lock washers. Remove spring plate or torque spring. Lower jack at side rail so that the spring is free from axle.

Remove pivot bolt nut and drive out pivot bolt from spring bracket and bushing Fig. 5.

Remove bushing from "U" shackle.

To install spring, replace pivot bolt first and then the "U" shackle bushing. Raise jack and place center bolt in spring saddle and install "U" bolts and nuts. "U" bolt nut, torque wrench reading, 50-55 ft. lbs., when torque reaction spring is used, 60-65 ft. lbs. Spring pivot bolt nut, 27-30 ft. lbs.

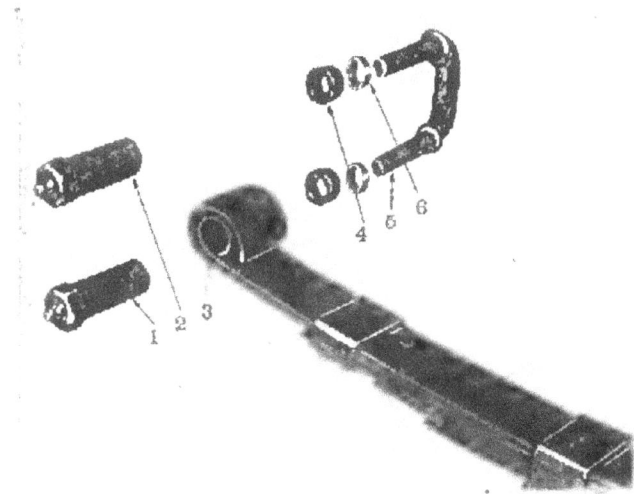

FIG. 3—SPRING SHACKLE—Left Front Spring

No.	Willys Part No.	Ford Part No.	Name
1	635532	GPW-5463	Spring Shackle Bushing Assembly (Left hand Thread)
2	634432	GPW-5464	Spring Shackle Bushing Assembly (Right hand Thread)
3	A-612	GPW-5311	Front Spring Assembly—Left
4	A-515	GPW-5481	Spring Shackle Grease Seal
5	A-513	GPW-5778	Spring Shackle U-Bolt (Left hand Thread)
6	A-1252	GPW-5482	Spring Shackle Grease Seal Retainer

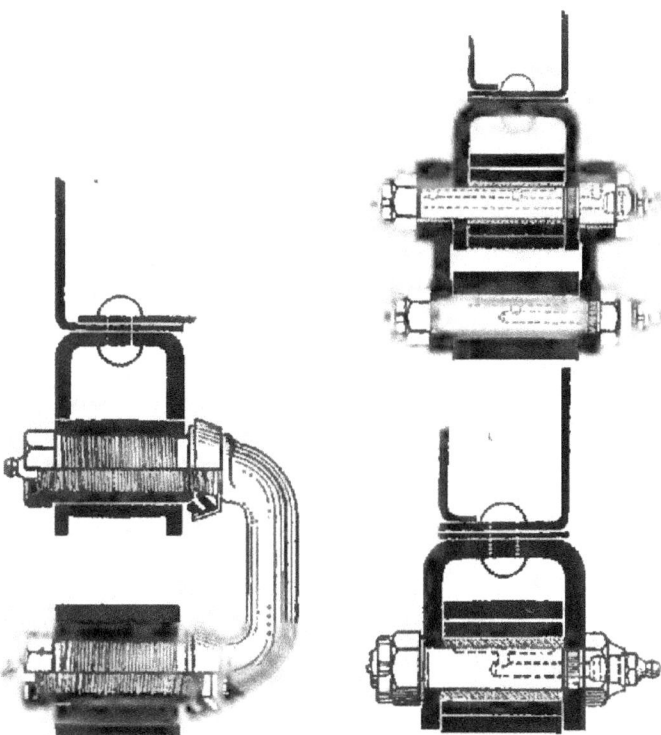

FIG. 4—SHACKLE & BOLT

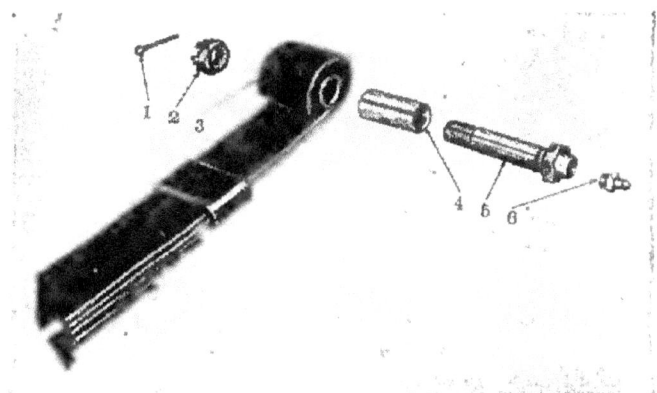

FIG. 5—SPRING BOLT—Right Rear Spring

No.	Willys Part No.	Ford Part No.	Name
1	5021	72034-S	Cotter Pin
2	6436	34033-S	Spring Bolt Nut
3	A-614	GPW-5560	Rear Spring Assembly—Right
4	359039	GPW-5781	Rear Spring Bolt Bushing
5	384228	GPW-5468	Spring Bolt
6	392909	353027-A1-S7-8	Grease Fittings

Shock Absorbers

The shock absorbers, Fig. 6 provide a much smoother ride by dampening the spring action as the vehicle passes over irregularities in the road.

The shock absorbers are the direct action type, two-way control and adjustable. The range of adjustment is four turns. To adjust the shock absorber, remove the lower end from the spring plate, push the unit together to engage the adjusting key and turn in a clockwise direction until the limit of the adjustment is reached. Holding adjusting key in slot, turn lower end anti-clockwise two turns. This is the average adjustment. Turning the adjustment to the right, or clockwise, gives a firmer control for rough roads, while turning in the opposite direction gives a softer control, allowing faster spring action. Should squeaks occur in the rubber mounting bushings, do not use mineral oil or rubber lubricant, but add a flat washer on the mounting pins to place the bushing under pressure and prevent movement between the rubber and the metal part.

To install shock absorbers, install inner mounting rubber bushing on both upper and lower bracket pins, install shock absorbers, install outer bushings, flat washer and then compress, inserting cotter key and spreading it to hold washer in proper position.

The shock absorber is sealed at the factory with the proper amount of fluid and is non-refillable.

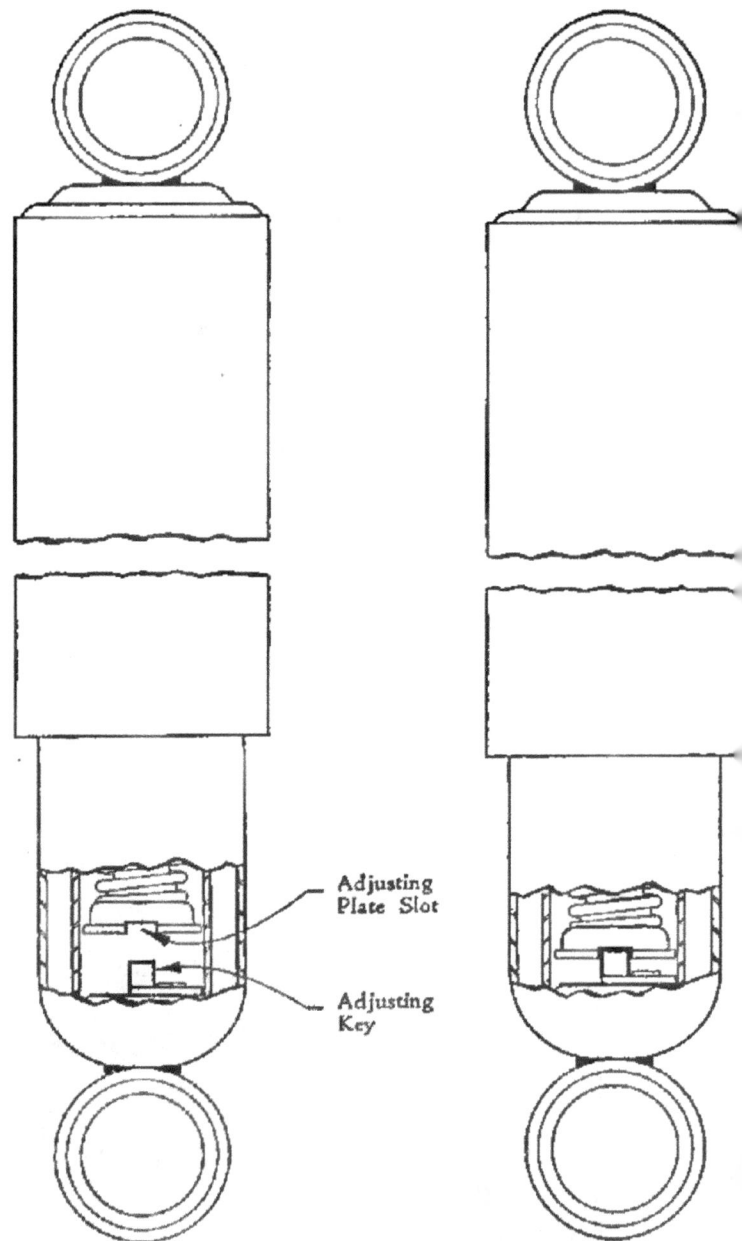

Adjusting Plate Slot

Adjusting Key

Sketch showing shock absorber before engaging adjusting slot and key.

Sketch showing shock absorber completely collapsed with adjusting key engaged in adjusting plate slot.

FIG. 6—SHOCK ABSORBER

SPRING TROUBLES AND CAUSES

SYMPTOMS	PROBABLE CAUSES
Spring Breakage—At center Bolt.............	Loose Spring to Axle Clips
Main Leaf Breakage on Ends...............	Tight Shackle or Pivot Bolt Shock Absorber Control Weak Poor Lubrication Spring Rebound too Great
Excessive Wear on Shackle Bushings........	Inside Spring Eye Opened Up Bushing Improperly Installed Lack of Lubrication Worn Bushings
Shock Absorber Noise....................	Lack of fluid Damaged Cylinder Loose Mounting Brackets Mounting rubber bushings worn out
Shock Absorber Control..................	Adjust Lack of Fluid—replace shocks

SPRING SPECIFICATIONS

Front Spring:

Make...........................	Mather
Type Leaf......................	Parabolic
Length Center to Center of Eye........	36¼"
Width..........................	1¾"
Number of Leaves...............	8
Front Eye Center to Center Bolt.......	18⅛"
Rear Eye Center to Center Bolt........	18⅛"
Left Camber under 525 lbs.............	5/16"
Right Camber under 390 lbs...........	5/16"
Rear Eye Bushed	
Bushing Size.........1¾" long I.D., .5655"	
Rebound Clips....................	4

Rear Spring:

Make...........................	Mather
Type Leaf......................	Parabolic
Length..........................	42"
Width..........................	1¾"
Number of Leaves...............	9
Rebound Clips...................	4
Camber under 800 lbs.............	¼"
Eye to Center Bolt...............	21"
Front Eye Bushed......1¾" long I.D., .5655"	

Shock Absorber Specifications

Make	Front-Monroe	Rear-Monroe
Type	Hydraulic	Hydraulic
Action	Double	Double
Length Compressed	10 9/16"	11 9/16"
Length Extended	16⅛"	18⅛"
Adjustable	Yes	Yes
Mountings	Rubber	Rubber

BODY

The body is of all-steel construction with mountings that provide a secure attachment to the frame.

All major panels are of No. 18 gauge steel. All open edges of panels are turned under, reinforced and flanged to give inherent strength. The panels are completely reinforced with "U" sections and welded. All component panels are seamed and welded together.

Body is insulated from frame with live rubber and fabric insulator shims placed between body and frame and held in place with body bolts.

The instruments and controls mounted on instrument panel are within clear view and easy reach.

Brass plugs have been placed in the left and right front corners to drain the floor.

Axe and Shovel

The axe or shovel can be removed or installed individually. The removal is apparent for they are held in place by fabric straps.

When installing the axe, turn the bit or blade up and place the handle in the front clamp, then insert the blade in the sheath after which pull up the clamp under the axe head and strap in place.

To install the shovel, turn the face against the cowl and place it in the strap on the cowl side. Next wrap the fabric strap through the handle and over the grip then between the grip and side of body; through the footman loop; over the outside of the grip and then buckle. This will hold the shovel forward securely in the strap on cowl side.

Windshield

The windshield and frame assembly provides for lowering entire assembly down on top of hood and also for opening and closing windshield for ventilation, when assembly is in upright position.

To lower windshield down on top of hood, loosen thumb screws on cowl so windshield can pivot and release two clamps holding tubular frame to instrument panel. When assembly is lowered onto hood fasten in place with hooks that are attached to the sides of hood.

To adjust windshield for ventilation, loosen the two wing nuts on upper brackets on each side of windshield, then entire frame assembly can be swung outward, anchor in position by tightening the wing nuts.

To Replace Windshield Glass

To replace windshield glass the following procedure should be followed:

1. Remove screws in each side of windshield adjusting bracket at top.
2. Bend down lip on left hand outer end of hinge at top.
3. Open windshield sufficiently to clear windshield frame and slide assembly off of hinge to left.
4. Remove the three nuts and bolts which hold the upper glass channel to frame.
5. Remove upper glass channel.
6. Withdraw glass from frame.

The replacement of windshield glass is the reverse of the above operations excepting that there should be new glazing tape used around the glass.

Windshield Seal

Windshield frame is sealed, when in closed position, by tension of special rubber seal against the tubular frame. The cowl seal, which is attached to the tubular frame is sealed when the clamps attaching frame to instrument panel are in locked position.

Top

To install the canvas top, it is necessary to loosen the two thumb screws at the pivot bracket, then slide tubing back out of front bracket, place in rear bracket and tighten thumb screws, allowing front bow to drop down over seats.

Attach canvas cover to top of windshield by the fasteners, then stretch canvas over bow and down to body back panel, placing the straps in the metal loops attached to body panel, stretch top and buckle straps. Next, raise front bow and assemble in the three bow flaps. The canvas top is carried under right front seat and held in place by straps.

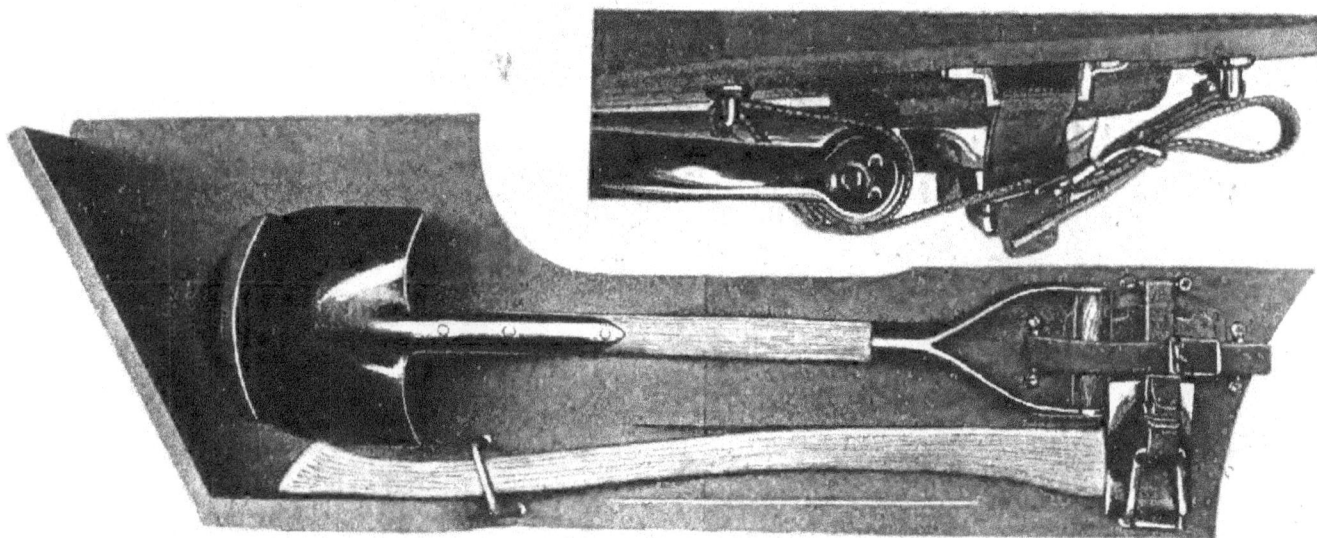

FIG. 1—SHOVEL AND AXE MOUNTING

TOOLS

Click here for Numerical Index of Tools

The manufacturers of the mechanical units used in this vehicle recommend the use of special precision tools, assembly jigs, gauges and close inspection of each part for assurance of proper operation and maximum service from each unit.

When necessary to perform a major operation on any mechanical unit special tools facilitate disassembling, checking and reassembling of the unit.

To aid the mechanic in performing satisfactory repairs, we suggest that tools as listed in this section or their equivalent be available when making major repairs.

OPERATING INSTRUCTIONS FOR SERVICE TOOLS

Supplied by Kent-Moore Organization
Detroit, Michigan

KMO-104—UNIVERSAL JOINT SNAP RING PLIERS. A special tool with jaws shaped to facilitate the removal and replacement of universal joint retainer rings.

J-270-1—DRIVER HANDLE. A heavy duty driving handle with a threaded end, on which can be mounted various adapters for removing and replacing bearing cups, oil seals, etc.

KMO-355—FEELER GAUGE SET. This feeler set consists of a number of blades, mounted in a suitable holder. The blades furnished total .040". The combination of blades provided herewith, will be found extremely useful in checking and reconditioning axle assemblies.

KMO-358—DRIVE PINION NUT WRENCH. This wrench is made to fit the retaining nuts that hold the several drive flanges to their splined shafts. The nuts are assembled very tightly and a heavy shank is necessary on the wrench in order that the nut can be loosened from its fit by a number of smart blows by a lead or copper hammer. This wrench consists of a 1¼" double broached hexagon socket, and a 15" hinged handle.

J-589-S—DRIVE PINION SETTING GAUGE SET—consists of:

1 J-589-1—Spindle and Micrometer Assembly
2 J-589-10-1—Locating Discs
1 J-589-H-1—Clamping Plate for Hypoid Attachment
1 J-589-H-5—Clamp Screw
1 J-589-10-2—Offset Plate for checking Hypoids
2 J-589-H-3—Hex Head Cap Screws
2 ⅝₆" Std. Plain Washers

1 J-589-SX—Master Gauge for Micrometer checking
1 J-589-B-1—Carrying Case

The rear axle pinion must be adjusted properly before any attempt is made to adjust the ring gear or differential. The ground face of the pinion is etched with its correct setting. The marking may be zero (0) minus (—) or plus (+). To determine the pinion setting, remove the differential and ring gear assembly. Bolt the clamping plate H-1 and screw H-5 across the open end of the axle housing in such manner that the hypoid (offset) plate, detail J-589-10-2 is clamped firmly to the end of the pinion. Next, place the locating discs, detail J-589-10-1 on each end of the micrometer spindle body and lower into position in the differential side bearing bores. Run the micrometer spindle down until the end contacts the free end of the hypoid offset plate, detail J-589-10-2. Rock the tool gently, adjusting the micrometer until it just drags slightly when rocked through a small arc, and note the reading on the micrometer spindle. A pinion marked "0" when properly adjusted, should show a micrometer reading of .719". A pinion marked "plus 2" when properly adjusted should show a micrometer reading of .717". A pinion marked (-4) should show a reading of .723" when properly adjusted. If pinion setting is not correct it will be necessary to remove the drive shaft flange, the pinion forward bearing cone, and the pinion, and remove or add shims as required between the rear bearing cup and the housing.

J-789—DRIVE PINION AXLE, AND TRANSFER UNIT FLANGE HOLDING TOOL. Used while removing the drive pinion flange to keep the pinion shaft from turning, and assists materially when removing or replacing the pinion shaft nut.

HM-872-S—DIFFERENTIAL SIDE BEAR-ING AND DRIVE PINION FLANGE RE-MOVER SET. This tool consists of a Puller Body, with an adapter plug HM-872-4 for use when removing side bearings, and a screw-end adapter sleeve HM-872-S-3 for use in removing Drive Pinion Flange. The fingers of the tool can be adjusted to the part being removed by means of the hinged yoke and thumb screw. Keep screw threads thoroughly lubricated during operation of the tool, and tap head of screw with a lead hammer in order to assist in removal. This same tool, when used with HM-872-S-4 Adapter, will remove the dust shields from Transfer Case Brake Drum Flange and Drive Flange.

J-881-A—UNIVERSAL JOINT ASSEMBLY AND DISASSEMBLY TOOL consists of Tool Assembly, composed of C-clamp, Screw and Swivel, and detail J-881-6 Cup for receiving the roller bearing assemblies of the universal joint trunnion, while removing. To use the tool as a replacer, place bearing and retainer assemblies on each end of pin and press together until the assembly can be placed in the joint. Leave tool in place until the U-clamps at each end of the trunnion are pulled up just tight enough to keep bearings from slipping out of place. Then remove tool and finish tightening U-bolts. When disassembling a joint place J-881-6 cup on lower plug of tool. Remove snap rings from their seats. Assemble tool to joint and turn down on screw until bearing retainer and needle rollers come free into the cup. Then reverse the tool, and press on end of pin until opposite bearing retainer and rollers drop into the cup. By using the tool in this manner the danger of losing rollers or dropping them is eliminated.

HM-914—REMOVER PLATES FOR RE MOVING BEARING CONE NEXT TO DRIVE PINION HEAD. The split halves of this tool are assembled to the drive pinion bearing, and bolted in place. By means of an arbor press, the pinion shaft can be pressed through the bearing. If an arbor press is not available, a suitable hand press can be devised by using Tool No. J-1759 with a pair of ⅜" x 8" standard thread bolts, that are furnished with the tool.

J-943—FRONT AXLE OUTER OIL SEAL REMOVER. The legs of this tool are made of tempered spring steel, and when compressed by hand, can be readily inserted behind the oil seal. A few taps by a hammer on the tie bar of the tool readily removes the seal.

SE-1066—RING GEAR BACK LASH CHECK-ING ATTACHMENTS—consists of:

   1 SE-1066-2—Clamp

   1 SE-1066-3—Connection for dial indicator with back mounting lugs

   1 SE-1066-1—Sleeve

These attachments allow a dial indicator to be set up so that the contact button of a dial indicator will come in contact with one of the gear teeth. As the ring gear is rocked back and forth by hand, the dial indicator will show the amount of back-lash between the ring gear and pinion.

SE-1094-5—DIAL INDICATOR. Not fur-nished with SE-1066 Attachment Set, but can be ordered extra if desired.

J-1375-S—FRONT PINION SHAFT FLANGE AND AXLE SHAFT FLANGE REPLACER. Under no circumstances should this flange be driven into place because of the possibility of damage to other parts of the unit. To operate place the flange in position on its shaft. Next, screw the socket adapter on end of pinion shaft, first placing the spacer washer between tool and end of shaft. Operating outer sleeve of the tool pushes the flange squarely and safely into place. Be sure the threads of tool are lubricated before each installation. Tool when used with adapters, S-1 and S-2 will also replace Transfer Case Brake Drum Flange and Main Transmission Drive Gear.

J-1436—WHEEL BEARING CUP AND OIL SEAL AND UTILITY PULLER. This item is a general utility tool with a wide range of uses such as removing oil seals, bearing cups, etc. Fingers are expanded or retracted by merely turning the handle right or left. A heavy sliding knocker that guides on the tool shaft and strikes against a lug welded to end of shaft, provides powerful leverage in removing parts pressed in place in various assemblies.

J-1735—FRONT AXLE SHAFT FLANGE PULLER. The puller body is so designed that its legs straddle the shaft and fit the under side of the flange. By turning down on the screw the flange is readily removed. In case of an exceptionally tight fit a few light blows with a lead hammer while the screw is under tension, will aid materially in the amount of effort required to free the flange.

J-1742—DRIVE PINION OIL SEAL RE-
MOVER. This tool was designed to remove the
oil seal without removing the pinion or differential.
Consists of a body with a center puller screw, and
four floating hooked fingers. To operate, turn
the fingers so they will slide through the opening
between the shaft and the oil seal, and when in
position, turn the finger ends into the locking slots
in the tool body. When fingers are in position the
striking sleeve is propelled against the head of the
center shaft until the oil seal is free of its seat in
the housing. Illustration shows a screw to push
against end of pinion shaft but this design has been
changed to a headed shaft with a sliding knocker.

J-1743—DRIVE PINION OIL SEAL AND
FRONT HUB OIL SEAL REPLACER. De-
signed to replace the oil seal without damage, and
with the pinion in place.

J-1744—FRONT AXLE WHEEL BEARING
ADJUSTING NUT WRENCH AND HANDLE.
This hollow wrench is designed with a pilot guide
ring on the inside of the body to prevent the
wrench from slipping off the thin adjusting nuts.
This construction permits tremendous pressure
being applied without danger of the wrench slipping off and injuring the operator.

J-1751—KING PIN BEARING CUP RE-
MOVER AND REPLACER AND FRONT AXLE
SHAFT OUTER OIL SEAL REPLACER. Con-
sists of a driver head and handle for removing the
bearing cup, and an adapter ring, J-1751-3,
for replacing the cups without damage. This tool
is also used in replacing the front axle shaft outer
oil seal.

J-1752—FRONT AND REAR AXLE SHAFT
INNER OIL SEAL REMOVER. Consists of a
screw with hinged ear, a cross bar and forcing nut.
This oil seal is readily removed by slipping the
hinged ear through the seal, and pulling it forward
into position. A cross-bar assembled over the
puller screw serves as a plate over the differential
bore. As the forcing nut is turned down against
the cross bar, the seal is removed from its position.

J-1753—FRONT AND REAR AXLE SHAFT
INNER OIL SEAL REPLACER. This is a special
driver designed to replace the oil seal without
damage. The tool has a short shank to enable it
to be operated in the confined area of the axle
housing. A short mallet must be used because of
the confined area in which to operate. See also
J-1753-3 Adapter.

J-1753-3—FEED SCREW ADAPTER FOR
USE WITH J-1753. In the event that a short
mallet is not available, the Axle Shaft Inner Oil
Seals can be installed in the following manner:—
Place an oil seal on J-1753 pilot. Slip the socket
end of feed screw over the shank of J-1753.
Assemble disc J-270-14 to the threaded end of
feedscrew. By turning the hexagon end of feedscrew

with an open end wrench, the oil seal is forced into
position, the thrust being taken by disc J-270-14
mounted in the opposite side of the differential
opening.

J-1761—DRIVE PINION BEARING CONE
REPLACER FOR CONE MOUNTED NEXT
TO PINION HEAD. Use of this tool prevents
the drive pinion bearing cone from being damaged
while being installed on the pinion shaft, and
eliminates the danger of scoring or shearing the
shaft or of chipping the cone.

J-1763—DIFFERENTIAL SIDE BEARING
CONE REPLACER HEAD. Designed to operate
with J-270-1 Handle. This tool pilots in the hole
in the differential case and is so designed that the
pressure is taken directly on the cone. The cone
can either be tapped or pressed into place, and
when properly used, the tool will eliminate all
danger of distorting the bearing roller cage.

J-1764—PAIR OF HOOKS FOR REMOVING
FRONT SPINDLE LOCK WASHERS. The lock
washer which is placed between the bearing adjusting nut and the lock nut has a tongued ear that
rides in the spindle keyway. Removal is sometimes
difficult because of housing interferences, and these
hooks will materially assist in withdrawing the
washer from the spindle.

J-1765—BRAKE ECCENTRIC ADJUSTING
TOOL. This tool has two rectangular slots to fit
the eccentric adjusting lugs on brake shoe anchor
pins. The tool is designed to operate with box
type wrenches such as are supplied with mechanics
hand tool sets.

KMO-410—TRANSFER CASE MAIN SHAFT
SNAP RING REMOVING PLIERS. These
pliers with knurled lugs allow the snap ring to be
expanded out of its groove in the shaft and readily
removed.

HM-872-S-4—ADAPTER—TRANSFER
CASE BRAKE DRUM FLANGE AND DRIVE
FLANGE DUST SHIELD REMOVER. This
adapter fits on the end of the main screw of
HM-872-S puller and allows dust shields to be
quickly removed.

J-1375-S-1—ADAPTER—TRANSFER
CASE MAIN DRIVE GEAR REPLACER. The
main drive gear should never be pounded into place
because of danger of damage to the internal
mechanism of the transmission. Use this adapter
in connection with the regular shaft of tool J-1375-S
to force the gear into place.

J-1375-S-2—ADAPTER—TRANSFER
CASE BRAKE DRUM ASSEMBLY REPLACER.
The brake drum assembly should never be pounded
into place because of almost certain damage to
connecting parts. Using this adapter in connection
with the regular shaft of tool J-1375-S, the drum
assembly is forced into place by screw feed safely
and quickly.

J-1375-6—ADAPTER SLEEVE. Required as a spacer sleeve when using either J-1375-S-2 or J-1375-S-1 in replacing brake drum and main drive gear.

J-1748—TRANSFER CASE MAIN SHAFT FRONT BEARING REPLACER. The nose of this tool is specially designed to allow the bearing to be installed without damage. A copper, lead, or rawhide mallet should be used when replacing.

J-1749—TRANSFER CASE MAIN SHAFT FRONT CONE REMOVER. In order to disassemble the main shaft to remove from the case, use this wedge shaped tool. Insert the tool between the front bearing and the gear, and tap the tool with a lead, copper, or rawhide mallet until the bearing is wedged off the shaft.

J-1754—TRANSFER CASE REPLACER FOR MAIN SHAFT BEARING CUPS, FRONT AND REAR. This tool is designed so that the cups are seated in exactly the proper distance in the bore. Strike shank end of tool with lead, copper, or rawhide mallet until the stop shoulder bottoms, when replacing front bearing cup. Rear bearing cup protrudes slightly on outside of case when properly seated.

J-1755—TRANSFER CASE FRONT BALL BEARING REPLACER. This tool was designed to drive the ball bearing into place in the front bearing cap. The driver end of the tool contacts the outer race only and will not injure the bearing while installing. Use a lead, copper, or rawhide mallet with the tool. This tool can also be used for wheel hub inner bearing cone and oil seal.

J-1756—TRANSFER CASE OIL SEAL REPLACER. A specially designed tool with a pilot head to hold the oil seal while starting and to drive the seal in place without damage. Used for both front and rear bearing caps. A lead, copper, or rawhide mallet should be used with this tool.

J-1757 — TRANSFER CASE SHIFTER SHAFT OIL SEAL ASSEMBLY TOOLS. Consists of J-1757-1 Driver and J-1757-2 Pilot. The tapered pilot is placed on the end of the shifter shaft and allows the oil seal to be expanded gradually and uniformly as it is driven into position. Without the use of this tool, serious damage could result to the seal. When assembling shifter shaft, the pilot is also useful.

J-1758—SPEEDOMETER GEAR BUSHING REPLACER. The speedometer gear shaft rides in a small hardened bushing that is difficult to replace unless this special pilot driver is used.

J-1759—TRANSFER CASE BRAKE DRUM AND MAIN DRIVE GEAR PULLER SET. Tool can also be used as a hand press in connection with HM-914 for removing axle drive pinion bearing cone.

Tool consists of J-1759-1 Body; J-1759-2 Main Puller Screw; J-1759-3 Pair of Socket studs and check nuts for use in removing brake drum; two ⅜" x 8" bolts for use with HM-914 in removing drive pinion bearing; two ⅜" x 1¾" screws, nuts, and washers for removing Bantam type brake drum; and one pair of J-1759-5 Fingers and Nuts for removing transmission main drive gear. The body of this tool is slotted at each end, and various adapters can be added to make it a utility puller with a wide range of application.

J-1766—TRANSFER CASE FRONT AND REAR BEARING CAP OIL SEAL REMOVER. These oil seals are mounted tight against a shoulder and this specially designed drift provides a quick means of removal.

J-1767—TRANSFER CASE MAIN SHAFT SNAP RING INSTALLING SET. Consists of J-1767-1 pusher tube and J-1767-2 tapered thimble. The thimble pilots on the end of the shaft. The snap ring is started on the small end of the tapered part of the pilot. The forcing sleeve is used to force the ring on the shaft and into its groove.

J-1768—TRANSFER CASE IDLER GEAR THRUST WASHER LOCATOR. There is a thrust washer on each side of the case and without this tool it is difficult to assemble the idler gear cluster and at the same time keep the thrust washers from slipping out of position. To operate, place a film of grease on each thrust washer to help hold washer to case. Next, start the cluster gear shaft into the case, and it will help hold the washer in position on that side of the case. Place tool J-1768 through hole on opposite side with small end of tool centered into thrust washer. Install the gear cluster, push shaft through gear cluster, and it will pick up the opposite thrust washer and at the same time push the locator tool out of the way.

J-1769—TRANSFER CASE IDLER GEAR SHAFT DRIFT AND SHIFT LEVER PIN INSTALLING PILOT. This tool has a dual use. It is useful for forcing the idler gear shaft out of position. The same tool assists materially to locate the shift levers and springs as a pilot, which is pushed out of the way as the shift lever pin is pushed into position.

J-1770—TRANSFER CASE FRONT AND REAR DRIVE FLANGE DUST SHIELD REPLACER. This tool is a pilot driver correctly machined to allow the dust shields to be replaced without damage. Use a lead, copper, or rawhide mallet with this tool. This tool can also be used to replace wheel hub oil seals.

J-1771—TRANSFER CASE FRONT BEARING CAP SNAP RING REMOVER SET. Tool consists of J-1771-SA-1 tool to force snap ring out of groove, and J-1771-SA-2 hook to assist detail 1. This snap ring fits in an internal groove and is exceedingly difficult to remove with ordinary tools. To operate, pry one end of the snap ring out of its groove far enough to enter hook. When one end of the snap ring is free, it can be grasped by a pair of pliers and pulled out of the cap.

J-1772—TRANSFER CASE FRONT BEARING CAP SNAP RING REPLACER SET. Tool consists of the following parts:—J-1772-SA-1 sub-assembly into which ring is loaded and expelled into its groove; J-1772-SA-2 loading driver; J-1772-5 tapered loading sleeve.

To operate, place J-1772-SA-1 bottom side up in a vise, or with the handle protruding through a hole in the work bench. Next, place J-1772-5 tapered loading sleeve with large opening uppermost over the recessed collar of J-1772-SA-1. Then drop the snap ring into loading sleeve and use J-1772-SA-2 loading driver to put the ring through the tapered hole and into the recessed counterbore in J-1772-SA-1. With the snap ring installed in J-1772-SA-1, place the tool in position in the transfer case housing. Hold the outer case of the tool tight against the housing and strike the floating plunger smartly with a lead, copper, or rawhide mallet to expel the snap ring into its seat.

J-1773—TRANSFER CASE FRONT BEARING CUP REMOVER. After the main shaft cone has been wedged forward on the shaft (see tool J-1749 instructions) it is necessary to push the bearing cup out of the case in order to remove the shaft. Slide gear toward inside of case and insert tool J-1773 between the gear and the bearing cup. By driving on end of shaft with rawhide mallet, the cup is forced out of the case.

J-1774—TRANSFER CASE SPEEDOMETER GEAR SHAFT BUSHING REMOVER. (Not illustrated) No special tool has been provided for this purpose. Should service be required, any standard "Easy-Out", or tapered type stud extractor with a capacity of ¼" diameter, can be used to extract the bushing.

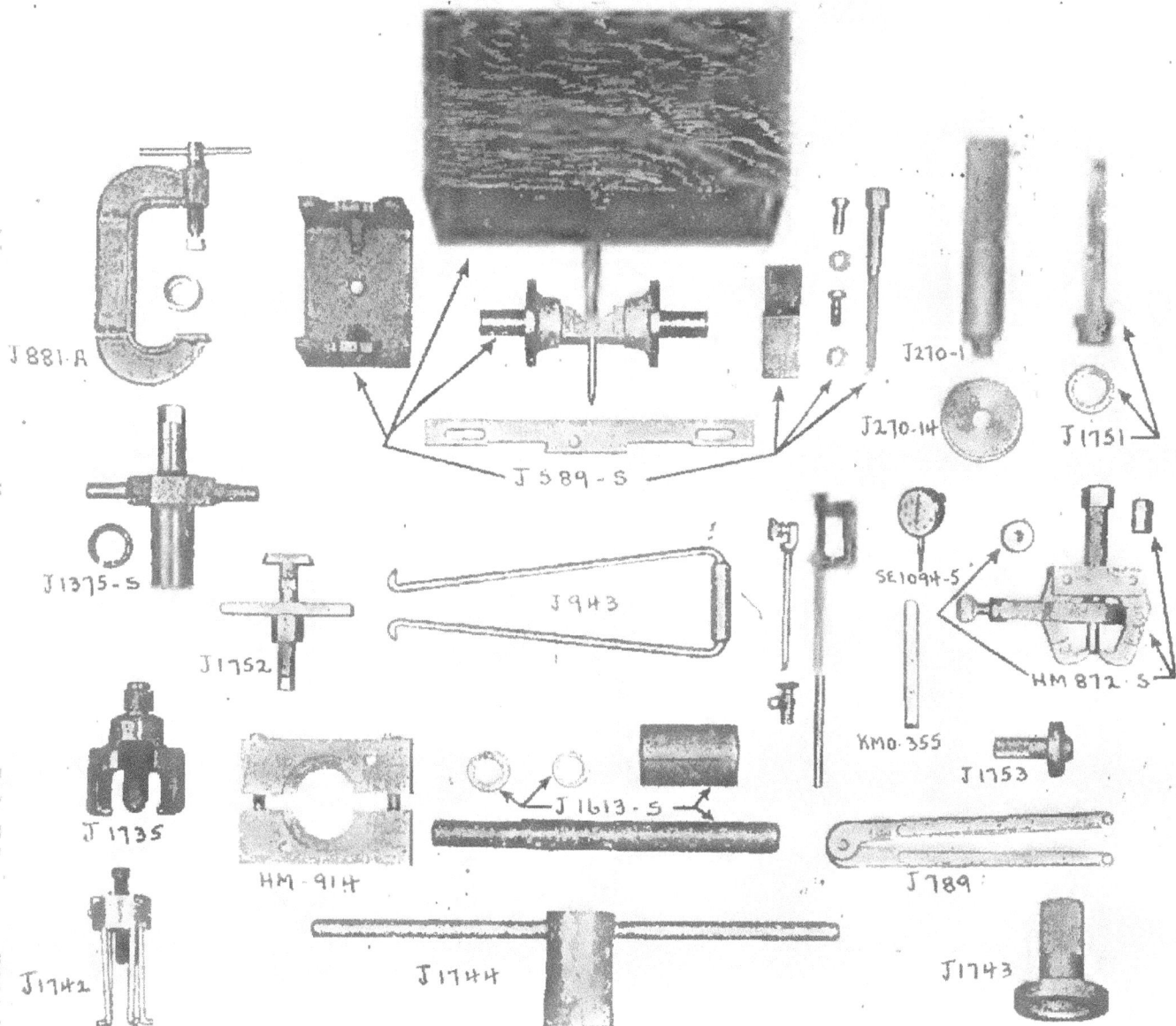

FIG. 1—SERVICE TOOLS

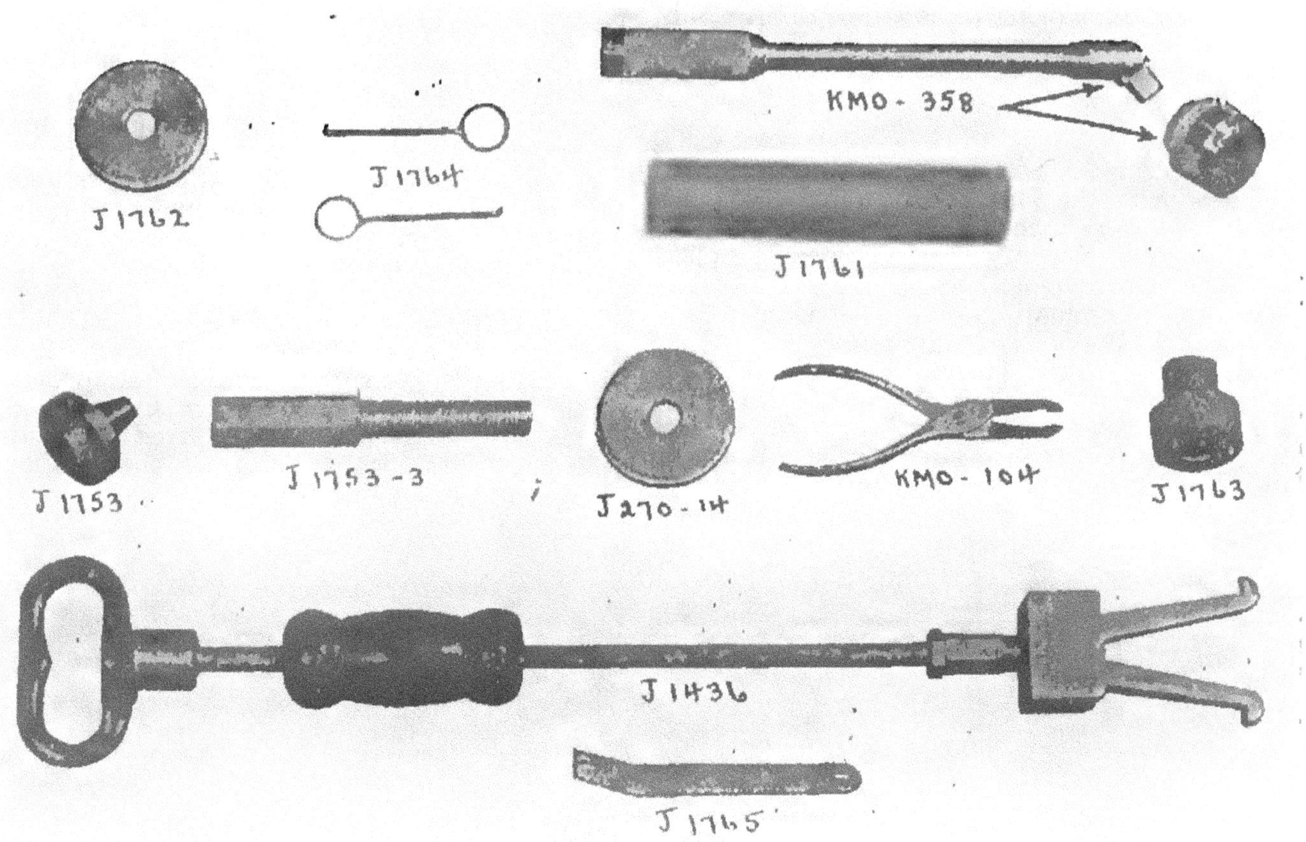

J1762

J1764

KMO-358

J1761

J1753

J1753-3

J270-14

KMO-104

J1763

J1436

J1765

FIG. 2—SERVICE TOOLS

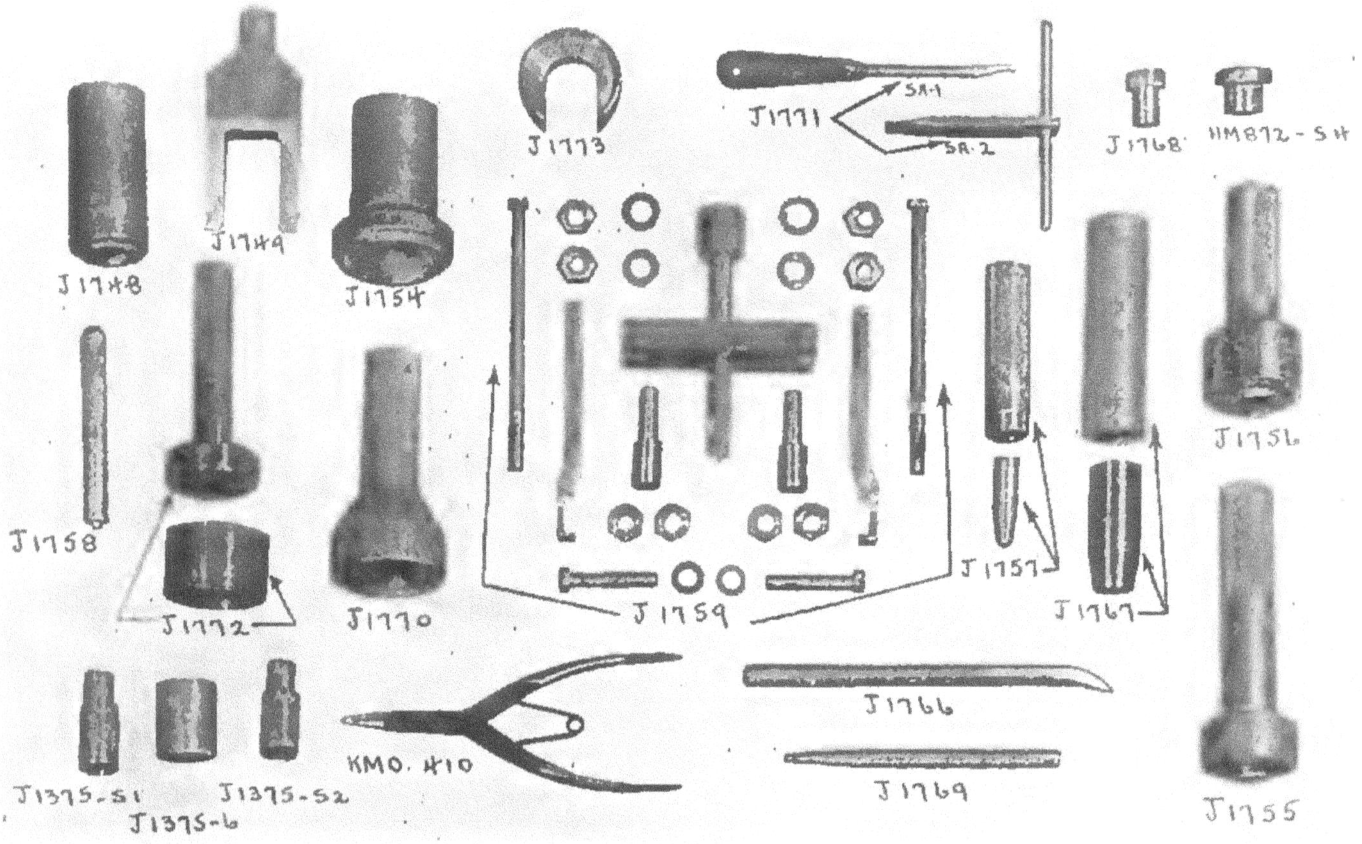

J1773

J1771 SA-1 SA-2

J1768 HM872-SH

J1748

J1749

J1754

J1756

J1758

J1757

J1767

J1772

J1770

J1759

J1375-S1 J1375-S2

J1375-6

KMO-410

J1766

J1769

J1755

FIG. 3—SERVICE TOOLS

NUMERICAL INDEX

SERVICE TOOLS REQUIRED FOR TRANSFER CASE

Supplied by Kent-Moore Organization, Detroit, Mich.

Willys No.	K-M No.	
A-6218	**KMO-410**	Transfer Case Main Shaft Snap Ring Removing Pliers..............
A-6200	HM-872-S-4	Adapter—Transfer Case Brake Drum Flange and Drive Flange Dust Shield Remover—works with HM-872-S ....
A-6201	J-1375-S-1	Adapter—Transfer Case Main Drive Gear Replacer. Operates with tool J-1375-S............................
A-6202	J-1375-S-2	Adapter—Transfer Case Brake Drum Assembly Replacer. Operates with tool J-1375-S............................
A-6203	J-1375-6	Sleeve required with J-1375-S-1 and J-1375-S-2 when replacing Transfer Case Main Drive Gear and Transfer Case Brake Drum............
A-6204	J-1748	Transfer Case Main Shaft Front Bearing Replacer................
A-6205	J-1749	Transfer Case Main Shaft Front Cone Remover...................
A-6206	J-1754	Transfer Case Replacer for Main Shaft Bearing Cups, Front and Rear.
A-6207	J-1755	Transfer Case Front Ball Bearing Cup Replacer.................
A-6208	J-1756	Transfer Case Bearing Cap Oil Seal Replacer...................
A-6209	J-1757	Transfer Case Oil Seal Assembly Tool Set for Shifter Shaft........
A-6210	J-1758	Speedometer Gear Bushing Replacer..........................
A-6211	J-1759	Transfer Case Brake Drum and Main Drive Gear Puller Set. Tool can also be used as a hand press in connection with HM-914 for removing Drive Pinion Bearing Cone...........................
A-6212	J-1766	Transfer Case Front and Rear Bearing Cap Oil Seal Remover........
A-6213	J-1767	Transfer Case Main Shaft Snap Ring Installing Set................
A-6214	J-1768	Transfer Case Idler Gear Thrust Washer Locator................
A-6215	J-1769	Transfer Case Idler Gear Shaft Drift and Shift Lever Pin Installing Pilot
A-6216	J-1770	Transfer Case Front and Rear Drive Flange Dust Shield Replacer.....
A-6217	J-1772	Transfer Case Front Bearing Cap Snap Ring Replacer Set..........

SERVICE TOOLS REQUIRED FOR FRONT AND REAR AXLES

Supplied by Kent-Moore Organization, Detroit, Mich.

Willys No.	K-M No.	
A-6243	**KMO-104**	Universal Joint Snap Ring Removing Pliers.....................
A-6221	**J-270-1**	Drive Handle..
A-6244	**KMO-355**	Feeler Gauge Set...
A-6245	**KMO-358**	Drive Pinion Nut Wrench Socket and Hinge Handle...............
A-6222	**J-589-S**	Drive Pinion Setting Gauge Set.............................
A-6223	**J-789**	Drive Pinion Axle and Transfer Unit Flange Holding Tool...........
A-6219	**HM-872-S**	Differential Side Bearing and Drive Pinion Flange Remover Set.......
A-6224	**J-881-A**	Universal Joint Assembly and Disassembly Tool.................
A-6220	**HM-914**	Remover Plates for Bearing Cone next to Drive Pinion Head........
A-6225	**J-943**	Front Axle Outer Oil Retainer Remover........................
A-6246	SE-1066	Ring Gear Back Lash Checking Attachments **less Dial Indicator or Feeler Gauge**..
A-6247	SE-1094-5	Dial Indicator..
A-6226	J-1375-S	Front Pinion Shaft Flange and Axle Shaft Replacer...............
A-6227	J-1436	Wheel Bearing Cup, Oil Seal and Utility Puller.................
A-6228	J-1735	Front Axle Shaft Flange Puller..............................
A-6229	J-1742	Drive Pinion Oil Seal Remover..............................
A-6230	J-1743	Drive Pinion Oil Seal and Front Hub Oil Seal Replacer.............
A-6231	J-1744	Front Axle Wheel Bearing Adjusting Nut Wrench and Handle........
A-6232	J-1751	King Pin Bearing Cup Remover and Replacer, Front Axle Shaft Outer Oil Seal Replacer, Transfer Case Output Shaft Bearing Remover....
A-6233	J-1752	Front and Rear Axle Shaft Inner Oil Seal Remover................
A-6234	J-1753	Front and Rear Axle Shaft Inner Oil Seal Replacer...............
A-6235	J-1753-3	Feed Screw for use with J-1753 Rear Axle Shaft Inner Oil Seal Replacer.
A-6236	J-1761	Drive Pinion Bearing Cone Replacer for Cone mounted next to Pinion Head..
A-6237	J-1763	Differential Side Bearing Cone Replacer Head (works with J-270-1 Handle)..
A-6238	J-1764	Pair of Hooks for Removing Front Spindle Lock Washer............
A-6239	J-1765	Brake Eccentric Adjusting Tool.............................
A-6240	J-1783	Differential Case Assembly Studs (Set of 4)...................
A-6241	J-1784	Drive Pinion Bearing Cup Replacer Set.:......................
A-6242	J-1785	Drive Pinion Bearing Cup Removing Set.......................

NUMERICAL INDEX

SERVICE TOOLS REQUIRED FOR ENGINE

Supplied by KENT-MOORE ORGANIZATION, Detroit, Michigan

Willys No.	K-M No.	
A-6248	KMO. 375 EX	Valve Guide Expansion Reamer—Size .375"
A-6249	KMO-213	Cylinder Compression Indicator
A-6250	J-1950	Valve Guide Removing and Installing Tool
A-6251	HM-593-0	Piston Fitting Gauge and Scale .003" x ¾" x 12"
A-6252	HM-593-10	Piston Fitting Gauge (Less Scale)
A-6253	KMO-402-B	Special ½" Tappet Wrench (Double End)
A-6254	KMO-402-BA	Special 17/32" Tappet Wrench (Double End)
A-6255	KMO. 812 Ex.	Piston Pin Reamer (Spec. Floating Pilot Expansion Type)—Size .8125".
A-6256	KMO. 913	Cylinder Bore Test Indicator
A-6257	J-1876-0	Connecting Rod and Piston Aligning Fixture
A-6258	J-1951	Piston Ring Installing and Removing Tool
A-6259	J-1952	Universal Clutch Shaft Pilot Arbor
A-6260	KMO-357	Universal Piston Ring Compressor
A-6262	J-1955	Valve Lifter
A-6261	KMO-144	Fuel Pump Checking Gauge & Vacuum Meter
A-6263	J-1953	Valve Lock Installing Tool
A-6264	J-1313	Tension Indicator Wrench Double Acting Beam Type.
A-6265	C 537	Voltmeter
A-6266	J-1954	Stud Remover and Installing Tool

INDEX

LUBRICATION SPECIFICATIONS
CANADIAN & BRITISH

ITEM TO BE LUBRICATED See Page 10 & 11	HOW APPLIED	Capac. Imper.	B.W.D. Specified	B.W.D. ☐ Emg'y	D.N.D. Summer	D.N.D. Winter	Nearest Commercial Equivalent Summer	Nearest Commercial Equivalent Winter	Miles
Engine Crankcase (39)	Filler pipe R. side check level daily	3½ qts.	M-120	M-160	51	45	Auto Eng. Oil—S.A.E. 20W	Auto Eng. Oil—S.A.E. 10W	2000
*Transmission Case (28)	Filler plug R. side—add Oil to level of plug	1¾ pts.	C-600	M-600	390	360	Truck Hypoid—S.A.E. 90	Truck Hypoid—S.A.E. 80	5000
Transfer Case (29)	Filler plug—add oil to level of plug	2½ pts.	M-800	C-600	140	100	Aviation Eng. Oil—No. 140	Aviation Eng. Oil—No. 100	5000
Differential F. & R. (30)	Filler plug in cover—add Hypoid oil to level of plug	2 pts.	Hypoid 90		390	360	Truck Hypoid—S.A.E. 90	Truck Hypoid—S.A.E. 80	5000
Propeller Shaft Universal Joints F & R (6)	Fitting	Fill	C-600	M-600	390	360	Truck Hypoid—S.A.E. 90	Truck Hypoid—S.A.E. 80	1000
Air Cleaner	Remove Cover	1 pt.	M-120	M-160	51	45	Auto Eng. Oil—S.A.E. 20W	Auto Eng. Oil—S.A.E. 10W	1000
Front Axle Shaft Universal Joint & Steering Knuckle Bearings (27)	Filler plug outer casing	½ lb.	Grease G.S.		632	632	Wheel Bearing Lub.	Wheel Bearing Lub.	1000
F & R Wheel Bearings (19)	Remove and Repack	Fill	Grease G.S.		632	632	Wheel Bearing Lub.	Wheel Bearing Lub.	5000
Steering Gear Housing (22)	Remove Plug	6½ oz.	C-600	M-600	390	360	Truck Hypoid—S.A.E. 90	Truck Hypoid—S.A.E. 80	1000
Steering Bell Crank (10)	Fitting		Grease G.S.		632	632	Chassis Grease	Chassis Grease	1000
Steering Tie Rods (3)	Fitting each end								
Steering Connecting Rod (4)	Fitting each end								
Spring Shackles F & R (1)	Fittings 8								
Spring Pivot Bolts F & R (2)	Fittings 5								
Clutch & Brake Pedal Shaft (10)	Fittings 2								
Propeller Shaft Slip Joints (7)	Fitting 1 each								
Starter Front (36)	Oil Hole	5 drops	M-120	M-160	51	45	Auto Eng. Oil—S.A.E. 20W	Auto Eng. Oil—S.A.E. 10W	1000
Distributor (34)	Oil Cup on side	5 drops	M-120	M-160	51	45	Auto Eng. Oil—S.A.E. 20W	Auto Eng. Oil—S.A.E. 10W	1000
Distributor Shaft Wick (34)	Oil Can	1 drop	M-120	M-160	51	45	Auto Eng. Oil—S.A.E. 20W	Auto Eng. Oil—S.A.E. 10W	2000
Distributor Arm Pivot (34)	Oil Can	1 drop	M-120	M-160	51	45	Auto Eng. Oil—S.A.E. 20W	Auto Eng. Oil—S.A.E. 10W	2000
Distributor Cam (34)	Wipe with grease		Grease G.S.		665	665	Mineral Jelly	Mineral Jelly	2000
Clevis Pins and Yokes (21)	Oil Can	5 drops	M-160	M-120	390	360	Truck Hypoid—S.A.E. 90	Truck Hypoid—S.A.E. 80	1000
Pintle Hook (40)	Oil Can	5 drops	M-160	M-120	390	360	Truck Hypoid—S.A.E. 90	Truck Hypoid—S.A.E. 90	1000

Hydraulic Brake Fluid No. 2 — 510 — 510 — Lockheed 21, Delco Super 9 or Iso

*Remove Skid Plate to drain Transmission
Shock Absorbers Non-Refillable
☐ Emergency

www.ingramcontent.com/pod-product-compliance
Lightning Source LLC
Chambersburg PA
CBHW080420030426